"SO I HEADED WEST"

(Ballarat to Broken Hill, to Kanowna, to Kalgoorlie)

By

W.G. Manners

1863 - 1924

(A Goldfields Pioneer)

W.G. MANNERS & CO
&
HESPERIAN PRESS

1992
W.G. Manners & Co
PO Box 33
45 Brookman Street
Kalgoorlie
Western Australia 6430

and

Hesperian Press
P.O. Box 317
Victoria Park 6100
65 Oats Street
Carlisle 6101
Western Australia

ISBN 0 85 905 123 4

Hesperian books are available direct from the publisher, but see your local bookshop first. For a full catalogue of Hesperian books on Australiana, natural history, exploration, Aboriginal ethnography, bush verse and gold prospecting please forward a self-addressed stamped envelope to the above address.
Cover design by Susan Ellvey of Designpoint, Perth

BACKGROUND

These are memories of early Ballarat/Broken Hill/Kanowna/Kalgoorlie by William George Manners (WGM) 1863 - 1924.

WGM's handwritten notes were compiled by his son Charles Brown Manners (CBM) during the late 1950's.

Assembly for publication with supporting material was commenced in 1988 by grandson, Ronald Brown Manners (RBM) 1936 - ? to mark the 90th Anniversary of WGM commencing his mining consultancy in 1898.

W. G. Manners & Co is still at Kalgoorlie (20km from Kanowna) and is managed by great grandsons, Craig, Scott and Ian, in partnership with their sister Sarah and Mannwest Pty Ltd.

Other related companies to emerge from the original W. G. Manners & Co nucleus are described on the wine bottle label reproduced below:

NOTES

1. The temptation to alter or update the text has been resisted, which explains why many early terms are retained i.e. "nigger", which in those days was commonly used in a "matter of fact" and even affectionate way, rather than any derogatory sense.
2. Government policies for inflating the currency have made it difficult to convert the figures quoted, into today's purchasing power. The Australian pound (£) was converted to $2.00 in 1966. The pound consisted of 20 shillings (20/-) of 12 pence (12d) each. The shilling ('Bob') was divided into sixpence (6d) and threepence (3d). As a guide, the male average weekly wage in 1912 was £3.
3. All measurements have been left as Imperial.

One inch	=	25.4	millimetres
One foot (12 inches)	=	304.8	millimetres
One yard (36 inches)	=	0.914	metre
One mile (1760 yards)	=	1.609	kilometres
An acre	=	0.405	hectare
One gallon (8 pints)	=	4.546	litres
One pound	=	.4536	kilogram
One hundredweight (112 pounds)	=	50.8023	kilogram
One ton (20 hundredweight)	=	1016.05	kilogram
One degree fahrenheit	=	.55	degree centigrade

Gold weighing is still carried out using troy weight.

24 grains	=	1	pennyweight (dwt.)
20 pennyweight	=	1	ounce troy
12 ounces	=	1 lb	troy
1 troy ounce	=	31.1034	grams.

GLOSSARY

grub-stake — to supply prospectors with provisions, equipment etc., in exchange for share of profits.

Hannans — old name for Kalgoorlie.

Navvy — labourer, usually employed in excavating.

Old Camp — former term for Coolgardie.

rise — passage being worked upwards from a level in a goldmine.

slug — a name often given by prospectors to a nugget.

stope — an underground excavation from which ore has been extracted.

tribute — the tributing system operated on the Golden Mile for many years. Companies allowed parties of miners to work privately on certain underground sections. This work was arranged under agreements, or tributes. The miners did not receive wages but got the gold they won, after paying the company for materials supplied, cost of treatment and a royalty on gold recovered.

weight — short for pennyweight, also referred to as dwt, a common term in goldmining prior to 1966.

winze — passage worked downwards in a goldmine, usually to test an ore body.

TABLE OF CONTENTS

ILLUSTRATIONS

PAGE

ABBREVIATIONS

EGHS	=	Eastern Goldfields Historical Society Inc.
MG	=	The Museum of the Goldfields - Kalgoorlie
Battye	=	Battye Library
BHS	=	Ballarat Historical Society

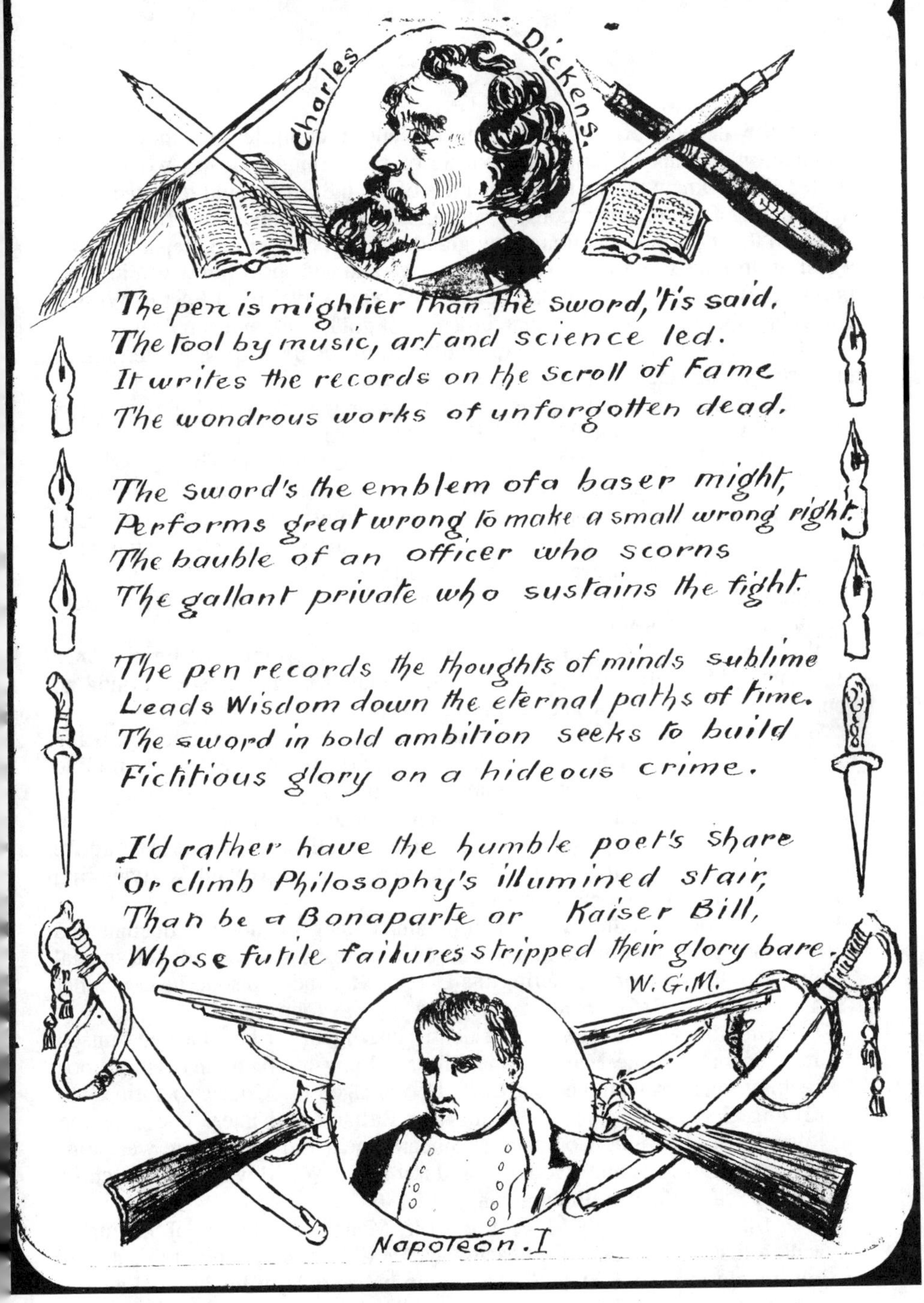

From a birthday card to daughter Muriel 16th September 1923

PREFACE

As the decade of the 1990's opens, and I compile this material for publication, I think of how the pioneers of our mining industry would have smiled if they knew the same uncertainties of the industry would repeatedly test the courage of future generations, in spite of all technical advances.

With this brief introduction to my grandfather's book let me prepare you for a visit to mining's humble beginnings in Australia and show how even in those pre-aircraft days the rat race had begun, involving travel from East to West in Australia, South Africa and London in search of gold, work and contracts.

My father has described my grandfather to me as a steadfast battler among the early mining legends and colorful entrepreneurs.

The son of a veteran of the Eureka rebellion he worked as a mining engineer for Claude de Bernales and future US President Herbert Hoover.

WGM's judgement of such people appears to be more on their successes for their shareholders, than their own fortunes. His comments on Horatio Bottomley establish that individual as a handy reference point at one end of the scale. Bottomley like that other arch-scoundrel of Australia's early gold mining days, Whittaker Wright, never even set foot on Australian soil.

Personally with Father Long at Kanowna, WGM now solves the mystery surrounding the often-told story of Father Long's "Sacred Nugget".

WGM's praise of Sir (later Lord) John Forrest and criticism of Jimmy (later Sir James) Mitchell only makes me wish that WGM was still around to comment on the current plague of "political fleas".

Born in 1863 WGM received a "kick start" into the mining industry, being the son of William Manners, then the manager of the Queen Victorian Gold Mining Company at Ballarat. William, a Scottish shipbuilder taking a job as ship's carpenter, was already hatching a plan to desert ship and "go for gold" at the exciting Ballarat diggings in 1853 when he fell in love with a youthful passenger to Australia (to become Mrs. Manners later that year) thus giving him another good reason to jump ship.

His transition from Ship's carpenter to mine manager is not well documented but his somewhat rebellious nature caused him to be singled out for special treatment by the authorities during the Eureka Stockade episode in December 1854. My father (CBM) records the next few years as follows:

> "Grandfather Manners was evidently intelligent for he educated himself and became a Mine Manager in Ballarat, doing his own surveying, book keeping, plan drawing and smelting. He managed the Queen Victoria Gold Mining Co. on the Canadian lead, near Ballarat and later, The Smeaton Reserve Gold Mining Company at Smeaton near Ballarat. There were nine children (four girls and five boys) and my father WGM was the sixth child, born at Scarsdale, near Ballarat, on 19.12.1863.
>
> At the age of 12 WGM went to a Mr. Menzies High School, having to walk a distance of four miles each way. He did not stay long at this school but transferred to the State School within half a mile of his home. Leaving school at the age of fifteen he was apprenticed to an engineering firm named Lonie & Dingle at 2/6d per week, but soon afterwards transferred to George Munro's Agricultural Implement Works at Alfredton at 5/- per week - after

eighteen months he again transferred to the Phoenix Foundry at Ballarat as an improver at 10/- per week, and had to walk two and a half miles each way for three days per week making the total walking ten miles per day on those three days. He continued at the Ballarat School of Mines for seven years and in 1886 received one of the first two Engineering Certificates ever issued by the Ballarat School of Mines.

In 1886 he went to the new Silver mines in Broken Hill where he worked on the original Broken Hill Proprietary Mine with Herman H. Schlapp, one of the Hill's leading Technical Pioneers. He then went as Foreman /Draftsman, graduating to Engineer of Block 14 Silver Mining Co., later transferring to the Tarrawingee Lime and Flux Company at Tarrawingee, the supplier to Broken Hill's smelters. That company having recently been formed by the legendary promoter James Smith Reid.

In 1889 he married Margaret Brown of Ballarat, in Ballarat, and returned to Broken Hill where they remained until 1895 (apart from 1891-93 when they were based in Tarrawingee).

He then went to Kanowna (near Kalgoorlie) in 1895. This was a new goldfield, discovered in 1893. The family followed in 1896. In 1900 they transferred to Kalgoorlie, where they completed their family of six children."

In later letters to my father (CBM), James Lamb (known as the "Prince of Prospectors") whose 1935 - 1938 letters are described at the conclusion of the MacDonnell Ranges chapter, refers to WGM as being the first mining engineer to visit Broken Hill and the Northern Territory. I have no way of verifying this, but as WGM visited the Broken Hill area before the sulphides were mined it is possible that Lamb's claims are true. WGM's description of the arrival of the first woman in Broken Hill (Chapter 7) adds weight to this claim.

Those "living" history, often fail to comprehend the significance of their times, but the passage of many years enable their strivings to be seen in true perspective.

Such was WGM's report of what was probably the first inspection and report on a mining prospect in the Northern Territory (MacDonnell Range Chapter). It is interesting to contrast his comments with the prospector/vendor's viewpoint, as recorded in James Lamb's letter. (By some strange coincidence, some ninety years later I found myself appointed Exploration Director for a company that owned the same Arltunga mining property that formed the basis of WGM's 1892 report).

In addition to WGM's engineering practice, he operated as Kalgoorlie's only Patent Attorney. His Patent Register, recording data of all patents lodged with him between 1901 - 1918 highlights the inventive nature of the goldfields diggers.

He was also a prolific correspondent to Technical Journals, with many published articles, some of which are included (in Appendix VI).

CBM intended to produce a limited edition of this book, for family circulation. This task was never completed by my father and now forty years later it appears to warrant a wider reading for what it tells of the people and life styles of those interesting times in Australia's formative mining history.

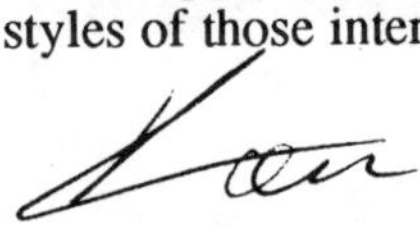

RON MANNERS (December 10, 1991)

WILLIAM GEORGE MANNERS
Born Scarsdale (near Ballarat)
19.12.1863 - 10.12.1924

INTRODUCTION

Why should I write the story of my life? I have done nothing of any consequence, have not made money or done any amount of good or harm; and, in fact, am only an ordinary obscure citizen of the State who has muddled through life, unknown, unhonoured, and unsung.

Yet perhaps there is something to learn from every person's life, no matter how obscure it may have been.

There are mistakes which, if told, may lead others to avoid the snares that the unscrupulous are daily laying to trap unsuspecting fools and inexperienced youths. There are things that one has learned by sad and bitter experiences which the rising generation have yet to learn: although that the rising generation will hardly learn from precept, but must gain the experience for itself. But there might be some who will benefit by what I write. That they will start their works from the points where the passing generation has left off, and so do some lasting good to the world. They would leave behind them at least "Some footprints on the sands of time"; or, start their life and works with such a handicap, that they may win some prizes in the race of life.

How often have I thought, if one could only start life with the knowledge and experience gained in sixty years struggle with the world, what wonders one might do, and what disastrous errors one might avoid! But that being impossible, one can do the next best thing, and pass on the experiences gained to those who may take note of the facts. This would give them the advantage which may save them a lot of preliminary toil and unfortunate side-steps.

With these thoughts in my mind, I propose to tell the simple story of my life: embracing its humours, its sorrows, and its mistakes, its successes and its failures; trying only to avoid the absolute commonplace events to dwell on those that will either amuse, educate, or throw some light on educational, professional, social or political questions.

It appears to me on looking back to my childhood days, that the overwhelming majority of children are good at heart. If they are given a proper chance the great majority will grow up to be men and women of good character and station.

I think there is some innate goodness in all of us that, if allowed to dominate, will guide us on the right track. I believe that there is a spirit within us that, if encouraged and allowed free play, will lead and influence our own advantage. This influence will always be good; it is only when we willingly flout its dictates that we fail or fall upon evil ways.

I am not building on the hope that these memoirs will ever reach the distinction of publication, but may be circulated among my own particular friends and descendants and perhaps their intimate friends. That even in this limited circulation it may sow some seeds that will propagate and multiply abundantly, ultimately leading to some good and useful work.

Then will my reward have been earned. For I hold that if we can do some good in the world, no matter how small that work may be, we have done our best. And what man can do more? I have desired all my life to do something worth doing; something that would add to the wealth or worth of the world. But circumstances have not aided me and my best ambitions have proved futile. I still, however, feel like Robert Burns when he said:-

That I for dear old Scotland's sake
Some useful book or work should make;
Or sing a song at least.

Note: This introduction and the following reminiscences were written in the latter part of 1924 just prior to the author passing away on the 10th of December of that year.

"Australia"

by W.G. Manners (written 1884)

Australia! Greatest Monarch of the Isles
Where fortune on her favoured people smiles,
Who, in their turn, to every vantage prone
May reap the harvest by some nature sown,

Dear hand by youth, of freedom and of health
Where none deny that Industry is wealth,
For there the Sun resplendently patrols
And sheds his radiance on ambitious souls.

Who onward strive and with requited toil
Unearth her riches, cultivate her soil,
Open her lands where Nature's secret stores
Expose abundance - Floral, Fruit and Ores.

Note: Other poems by W.G.M. have been collected and included in Appendices I & II

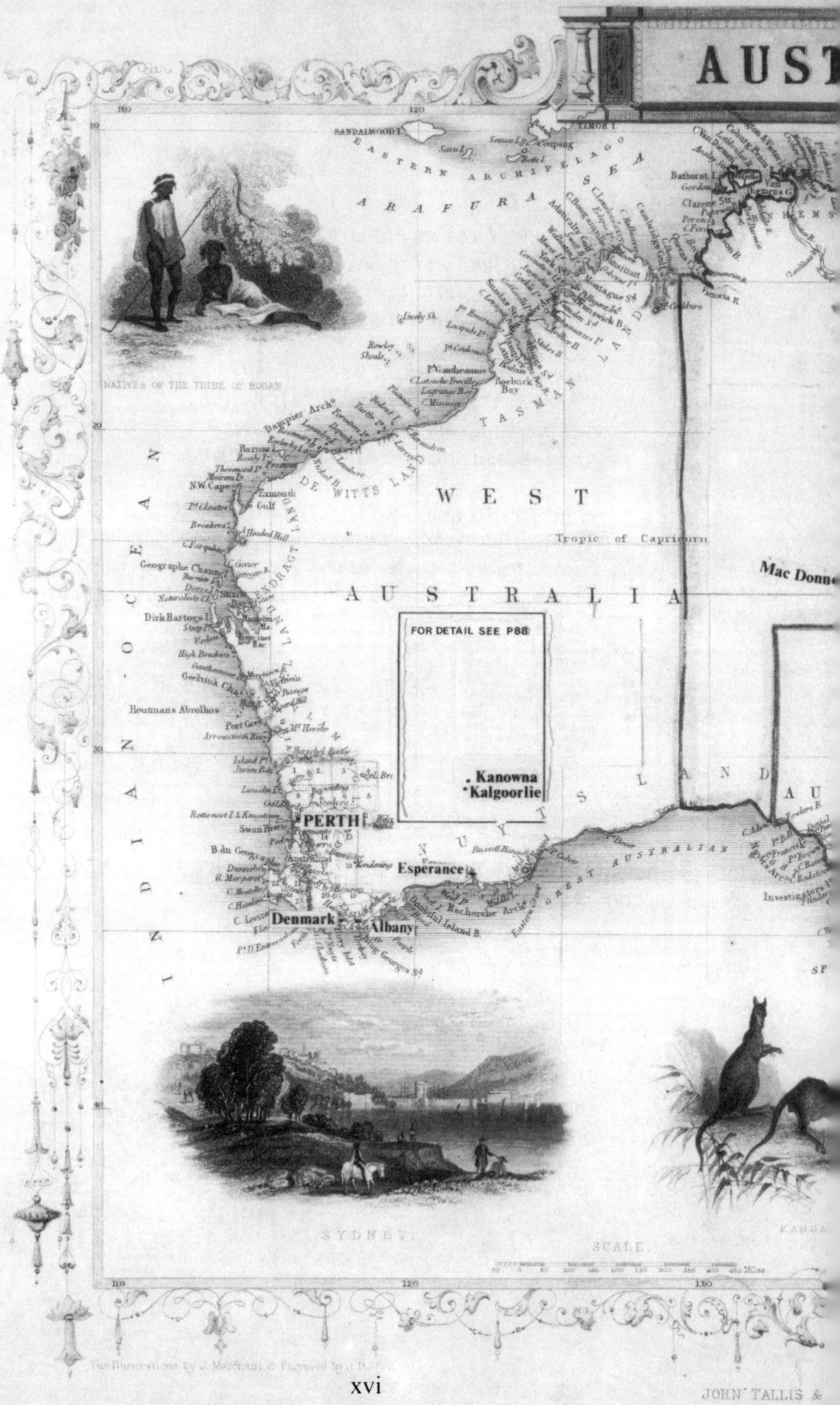
AUST
ARAFURA SEA
EASTERN ARCHIPELAGO
SANDALWOOD I.
TIMOR I.
INDIAN OCEAN
WEST AUSTRALIA
DE WITTS LAND
TASMAN LAND
EENDRAGT LAND
NUYTS LAND
GREAT AUSTRALIAN
Tropic of Capricorn
FOR DETAIL SEE P88
Kanowna
Kalgoorlie
PERTH
Esperance
Denmark
Albany
Mac Donne
Bathurst I.
Roebuck Bay
Dampier Arch.
N.W. Cape
Exmouth Gulf
Geographe Channel
Dirk Hartogs I.
Houtmans Abrolhos
Geelvink Chan.
Port Grey
Swan River
Recherche Arch.
NATIVES OF THE TRIBE OF BOGAN
SYDNEY
SCALE
JOHN TALLIS &

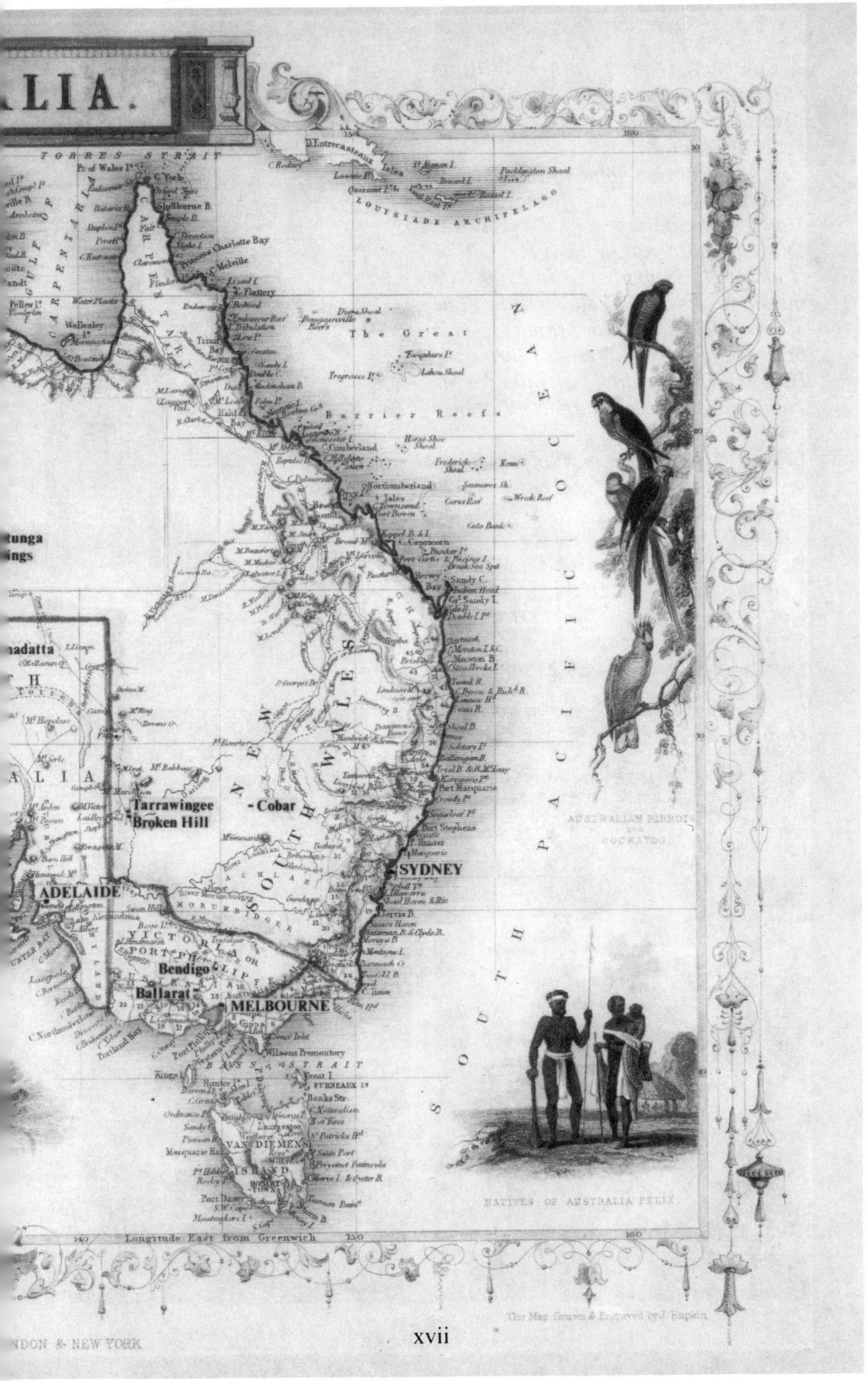
ALIA.
TORRES STRAIT
C. York
Shelburne B.
Princess Charlotte Bay
C. Melville
C. Flattery
LOUISIADE ARCHIPELAGO
D. Entrecasteaux Isles
Packington Shoal
The Great Barrier Reefs
Cumberland
Northumberland Isles
Port Bowen
Keppel B.
C. Capricorn
Port Curtis
Hervey Bay
Sandy C.
Moreton B.
Brisbane
Port Macquarie
Port Stephens
SYDNEY
NEW SOUTH WALES
Cobar
Tarrawingee
Broken Hill
ADELAIDE
Lake Alexandrina
VICTORIA OR PORT PHILLIP
Bendigo
Ballarat
MELBOURNE
Portland Bay
Port Phillip
Wilsons Promontory
BASS STRAIT
Kings I.
FURNEAUX Is.
VAN DIEMENS ISLAND
Macquarie Har.
Launceston
Hobart
Port Davey
SOUTH PACIFIC OCEAN
AUSTRALIAN PARROTS AND COCKATOO
NATIVES OF AUSTRALIA FELIX.
Longitude East from Greenwich
The Map Drawn & Engraved by J. Rapkin
NDON & NEW YORK

Right - *W.G.M.'s father William Manners jumped ship (the Ivanhoe) on 15th September 1853 and "went for gold" at Ballarat.*
He sailed from his home in Dumbarton, Scotland, where his father, John Manners was a Burgess (Alderman)
The romantic novel "Sweet Doll of Haddon Hall" published in 1909 tells the story of the 1567 elopement of one of his forebears, John Manners, with bridesmaid Dorothy Vernon, during the Haddon Hall wedding festivities for Sir Thomas Stanley and Margaret Vernon.

Lower Right - *William Manners later, as Manager of the Queen Victoria Mine, Ballarat.*

Below - *The other reason for William "jumping ship"; fellow passenger, Catherine Clark from Montrose, Scotland, who became Mrs Manners shortly after arriving in Australia.*

1

Childhood's Happy Days (1871)

"Ho-ki-la Ah Sam", said my brother Charlie as the Chinaman approached with a smile. He was carrying a wooden tub, consisting of half a forty gallon beer cask, on his shoulder, supported by one arm. We, my elder brother Alex, my younger brother Charlie, and I, had our boots and socks off. And had been catching crayfish in the dam alongside the main road, and having caught quite a number which we had in tin cans beside our boots on the bank of the dam, were playing and racing on the roadside when Ah Sam came along. He stopped among us and good naturedly greeted us. Charlie, who was always mischievous, could not resist the temptation to tickle him under his raised arm. This caused Ah Sam with a loud "Muck-a-hi" to quickly lower his arm with the result that the tub fell to the ground and, in the twinkling of an eye, was simply a confused heap of staves and hoops.

Before Ah Sam had fully realized the extent of his disaster, Charlie placed quite a stretch of the adjoining landscape between himself and the astonished Chow.

Ah Sam then made a rush to secure possession of the boots when we intervened, remarking, "You no take our boots, you takee his." "All ligh, you takem you boots, me takem his." agreed Ah Sam. Without any preliminary sign or word, Alex and I grabbed the three pairs of boots and were out of reach, leaving Ah Sam to swear in Chinese and make the best of it.

However, he was not completely unrewarded, as he collected our catch of crayfish and, gathering up the remnants of his tub, moved on his way mumbling to himself and shaking his fist at us, no doubt vowing vengeance on our heads if he ever caught us. But to do so he would have to have been very fleet of foot.

At this period of my career I would be about ten or eleven. My two brothers were within two years older and younger. I start the story at this period because it is the age in which are crowded many incidents, some amusing, but more bearing on those topics which I realize now have the greatest influence in forming the future life and character of the individual, viz. companionship and environment.

My father was a mine manager at the time, managing the Queen Victoria G.M. Co. on Canadian Hill, near Ballarat. He was a Scotsman who had left school when he was eight, then served his apprenticeship as a shipbuilder on the Clyde. These were, of course, the days of wooden ships. He had embarked as a ship's carpenter on a ship bound for Australia and landed in 1853. He deserted his ship, as sailors did in those alluring gold rush days, and had gone straight up to the diggings.

He was evidently intelligent; had educated himself and did all his own surveying. bookkeeping, plan drawing, and smelting.

He was just, but firm,kindhearted and generous, and I never remember him punishing any of us unless we thoroughly deserved it. And then it always did us good.

In the event of any accusation being made against us at any time, he would hold a kind of court martial, and elicit all the facts for and against. We would be punished or acquitted according to the evidence. We thoroughly realized the justice of this procedure and submitted to its dictates with good will.

On the other hand, my mother was entirely different. She made up her mind that we were guilty of a certain offence and, no matter what we said to the contrary, she stuck to her point. We had no redress. And, although I never told my parents a lie, I was so frequently contradicted by my mother that I seriously questioned the wisdom of telling the truth.

Such treatment is liable to make children argue that it is not worth while telling the truth if one is not believed when they do so. I think now that it was the even handed justice of my father that compensated for this failing in my mother which thus saved us from becoming liars of the highest order.

An illustration of this treatment occurred a little later in life, when I was working. I had at this time to rise at about 5.30 a.m., light the fire and boil the kettle, the table being laid for breakfast, and walk or mostly run about two and a half miles to the shop, work eight hours, then walk home again. Then, after tea, twice a week, walk the same distance to and from the School of Mines. I was also reading Dickens and, being deeply interested, would read as long as I could keep my eyes open. So I was often very sleepy and, in winter, it was quite dark when I had to get up.

There was a clock on a shelf beside the bed but I would have to strike a match to see the time.

My two younger brothers slept in the same bed. One morning I woke, struck a match and looked at the time. It was half past four. I threw the match away without blowing it out and immediately fell asleep again.

I was awakened some time later by my youngest brother calling out "Fire" I found the room full of smoke. On jumping out of bed I found that a large hole was burned through all the bedclothes. My match had evidently fallen short and remained alight. My youngest brother and I soon got water and extinguished the flames. One of the characteristics of my younger brother was that he remained in bed despite the smoke and subsequent water. He was of such a disposition that it required some violent action, such as an earthquake, to turn him out of bed one minute before it was absolutely necessary.

It being nearly rising time, I got my breakfast then went into my mother's bedroom and told her exactly what had happened.

When I came home from work, I met with this greeting, "This is a nice state of affairs, smoking in your bedroom and setting the bed on fire." I replied "I was not smoking I have told you the truth." "You were smoking. I know you were smoking" was my mother's reply. "Then, mother, you call me a liar." "No I don't, but you are not telling the truth now."

This hurt me more than I can tell and I never forgot it. Yes, what is the good of telling the truth?

My mother was also Scottish, from the east side of Scotland. She came out on the same ship as my father, with her mother, father and two sisters. She was, however, one of those Scottish people who have helped to give that excellent and generous-hearted people the reputation for being "Canny". She was the antithesis of my dad in this respect for whereas he was over-generous with his money, especially when he was on a bit of a spree, which often happened, she grabbed every penny and kept us all extremely short of cash.

We boys would often fossick or speck for gold. We often were successful in getting small quantities to the value of three or four shillings. Mother would take this from us and give us a penny or perhaps a half-penny. We soon learned, however, to keep our gold secreted and sell it to the gold buyers, from whom we received somewhere near its correct value.

Irrespective of these different temperaments in my parents, they both held that pernicious old Scottish notion that little boys should be seen and not heard. And as a consequence, when any elder person came to the house, we were relegated to the back yard to play. And we had our meals served to us in the kitchen.

Its effect was to make us bashful and ridiculously afraid of a top hat; I had grown well into manhood before I was able to meet a gentleman without a blush, or conduct myself like a well brought up young man should do.

Another disability that Scottish canniness imposed upon us was moleskin pants. They were cheap and strong in more senses than one. They wore for an interminable time but, no matter how long they wore, their aromatic strength was a menace to the neighbourhood. They possessed an aroma that spread in the atmosphere so that, as we approached any neighbourhood, long before we were seen, the folk with keen senses of smell would remark, "Here comes the Manners boys".

The advice is no longer necessary as regards moleskins I think, but oh, parents, dress your kids decently. Go without some finery yourselves if necessary, but make your children feel that they are not paupers. Give them a chance to acquire a modicum of self-respect. It will help them with their future careers and enable them to feel that they are as good as their neighbours. It will perhaps be the means of raising them from the mediocre to the eminent and worthy places in the world.

Living was exceedingly cheap in those days, and although wages were low, I am sure that the ratio of the cost of living to the wages was much more favorable in those days than it is at present.

All the restrictions on trade introduced both by the monopolists and the trade unionists have tended to raise the cost of living at a much greater rate than it has raised wages. And of these trade restrictions, I believe the trade union go-slow policy is the most iniquitous, the policy of protection the next.

In those days wages for labourers were about 30/- per week, miners £2/2/0, and tradesmen about £3 per week.

My father was a highly paid man at £6 per week, and I knew a Mine manager who considered himself a king among mining men at a "huge" salary of £8 per week.

However, one can appreciate the buying power of these wages when I tell you that my brothers and I would go up town with my mother on a Saturday night, taking a large cane perambulator with us to buy a side of mutton for half-a-crown, a barracuda "a yard long and a foot through, for a bob", and about 20 or 24 lbs. of fruit for a shilling, other things, of course being in proportion.

We would go to a market garden and after eating as much fruit as we could stuff, carry home as much rhubarb as our little backs were capable of sustaining,—all for an outlay of one shilling.

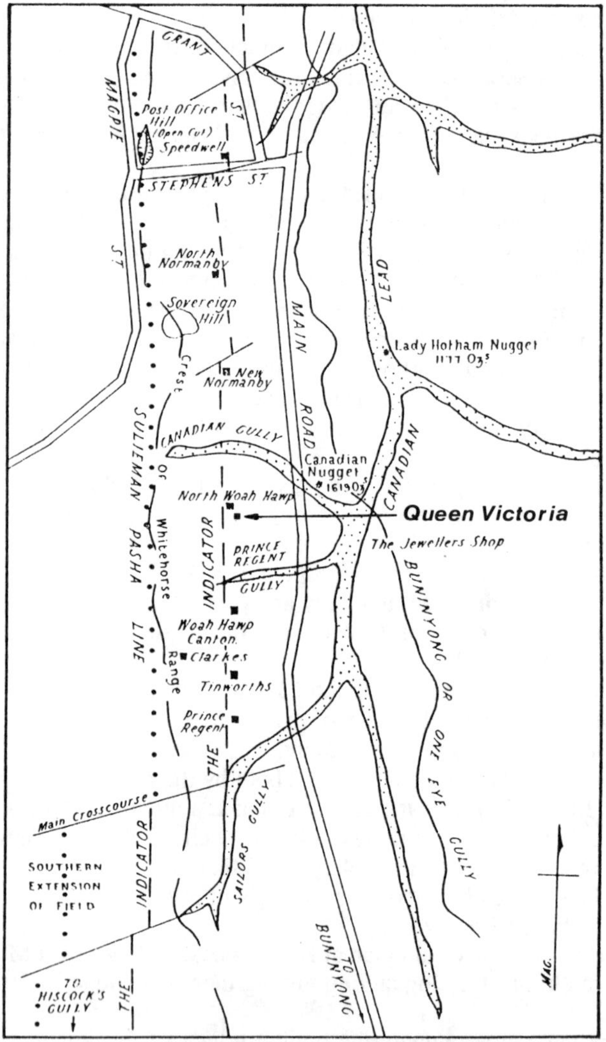

SOUTHERN PORTION OF
BALLARAT EAST FIELD
SHOWING POSITIONS OF
MAIN QUARTZ MINES
AND DEEP LEADS
(AFTER BARAGWANATH)

Scale of Feet
1000 500 0 1000

2

Mick And Tim (1877)

Our chief companions at this time were two step-brothers named Tom Melson and Tom Pardiner.

As they were both Toms, they were nicknamed respectively Mick and Tim, to avoid confusion.

They were both several years older than ourselves. Both were rather on the lower side of the moral standard of society.

They successfully taught us to swear, smoke, and chew tobacco, and also all the immoral side of sex matters. Thus at a very early age we learned these things, not in their religious and heaven endowed beauty, but as fun and a joke; all these vices were shown to us as being manly and clever.

Why we, who should have been brought up in quite a different social circle, should have been allowed to make this riff-raff our companions at this, the most dangerous and impressionable age of our lives, beats me to this day. But such it was, and I fancy that the same spirit which sanctioned moleskin pants allowed this anomaly.

Tim Pardiner was if anything the better of these brothers; Mick Melson was a big, hulking, foul-mouthed, cowardly fellow.

We, young as we were, soon recognised and took advantage of his cowardly disposition. On many occasions, when we made a fire on the hillside at night and sat around it to roast potatoes, that we would steal from our mother's larder, we would delight in telling ghost yarns. The surrounding darkness assisted to make these the more weird. Poor Mick would soon be trembling with fear and looking around him, evidently expecting to see a ghost appear out of the night. And when it came time to go home, a period which was generally announced by Mrs Melson, Tim's mother, calling "Mick, Tim," we would have to escort Mick to his very door.

One night he went home without his escort.

It was a Guy Fawkes night. We had pulled branches from the bush during the last three months, and built a splendid bonfire. We had duly incinerated our effigy of Guy, inundated him with fireworks, and were sitting around the dying embers, roasting our usual supper in the ashes when, out of the darkness, appeared a figure in white. In the best imitation of sepulchral tones that a child could supply it remarked, "Who dares burn my body?" Mick, who was not in on the secret, rose in a flash and made for home as fast as his bulk would permit. We were not able to induce him to venture out after dark for quite a fortnight.

Another occasion on which we scored rather an enjoyable joke off Mick was also a dark night. We were passing a Chinaman's house where six chows lived. Mick, who was always the first to originate mischief, threw some stones on the roof of the house which, being corrugated iron, made a great noise.

We were taken by surprise and, of course, Mick was twenty yards away before we got a start. We, however, scattered and ran for our lives and soon realized that the Chinamen had lost sight of us and were not following. But we saw Mick at top speed in the distance and set after him, as we ran muttering in gibberish as near as we could mimic Chinese. We kept a respectful distance behind Mick and he, thinking the Chows were behind him, made the pace pretty hot.

We thus chased him for about two miles, when Mick collapsed from sheer exhaustion. When we closed on him he was on his knees roaring, "You no touchee me. Me no do it!" when our vociferous mirth made him aware of the fact that we had put up a joke on him.

These fellows were our companions for about five years, after which we went to work and became acquainted with other and better companions.

We were afterwards stigmatised as stuck up and snobbish when we cut these acquaintances, but this made no difference to our change of policy.

3

My First Thoughts On Medical Efficiency (1880)

Our home on Canadian Hill was on the Queen Victoria Lease. The mine had been worked out. My father had bought the property and its lease was being worked by tributers.

The mine had always been low grade, but paid dividends of 1/- per month on 24,000 shares, with a return of 4 dwts. to the ton. (6 gms per tonne) But when the grade of ore fell much below this low value, operations on the mine ceased with, the above results.

There was a miner named Grant, a neighbour of ours, who had made a few thousand pounds in one of these tributes. One day, having charged a hole with blasting powder which was the principal explosive used at that time, a premature explosion took place while Grant was still in the drive.

He was not hit by any of the flying debris, but the full blast of air caused by the explosion in the confined space, struck Mr. Grant full in the face and knocked him down.

The only injury was to his eyes. They became very much inflamed and painful, so that the poor old chap, he was about fifty, had to sit in a dark room with a green shade over them.

Mr. Grant was not blinded, but the pain was severe, and he could not bear the light.

He consulted several eminent doctors in Ballarat without obtaining any relief. He then went to Melbourne and consulted some eye specialists, but all to no purpose.

After spending several hundred of pounds on the medical faculty, Mr. Grant was no better than before in fact, if anything, he was becoming worse.

At this time three American Specialists visited Ballarat, and advertised very largely. One was an eye specialist, another an ear, and the third a general practitioner.

Of course, here was Mr Grant's opportunity. What Australia could not do, the Great Yankee land might. So he went to see the eye specialist.

This magnate examined his eyes and assured him "I guess I can cure you in double quick time, and it will cost you only twenty pounds."

Mr Grant who had had considerable experience of medical methods asked. "If I give you twenty pounds are you quite sure that you can cure me?"

"Of course I can", replied the Yank.

"Then", said Mr Grant, "if you cure me I will give you twenty pound. If you don't I will give you nothing."

"Oh, no." remarked the quack, "We don't do business that way. We want cash down before we start."

"Then" said Grant, "I will give you fifty pounds if you cure me and nothing if you don't."

"No." said the Yank, "I have stated my terms."

"Well, I will give you a hundred pounds on the same conditions." suggested

Grant.

"No." was again the reply. Then Mr Grant boiled over.

"You are just a damned, defrauding quack like the rest of them." he said, and left the Yank to think it over at his leisure.

It was somewhere during this period that my Grandmother took ill. She was then a widow about seventy four years of age, undoubtedly very low and evidently becoming worse.

She was being attended by one of the best doctors in Ballarat. He could make nothing of her complaint, but was, of course, giving her medicine.

I suspect that this medicine was being given to her on what I afterwards devined was "diagnosis by elimination." That is to say, a doctor does not know what your complaints, but imagines it might be so and so. Then he prescribes some drug which would have a certain effect under these circumstances. If such an effect does not take place he tries some other drug to produce some other effect. And so he goes on eliminating certain diseases until he arrives at what he considers the right diagnosis; if the poor patient has the good fortune or rather the iron constitution to live through the process, he or she has a chance to recover by natural recuperation and in spite of the treatment.

My grandmother had an excellent constitution. Finally, the eminent physician called in two other doctors to consult with him as to the cause of this most mysterious illness. Their several and considered opinions were that the old woman was dying from old age, senile decay, a general break up of the system and that they could do nothing further for her.

This was very sad and consequently, I was dispatched to bring the Minister, the Reverand J.W. Inglis, who came down. We had a very solemn and tearful time.

Reverend Inglis prayed in a most effective manner, his words of comfort were eloquent and appropriate, and we were all in tears. While the old lady, who was a good Christian, was quite prepared to go. We all bade her an affectionate good-bye and expected never to see her alive again.

Of course the medicine was stopped because it was no use trying to prolong her life after such an impressive and important medical verdict.

But imagine our surprise when we learned of her gradual recovery. And, in a few weeks, she was almost as well as ever.

The dear old lady lived for twenty years after this, and died at the ripe old age of ninety five never having required the services of a doctor again.

This set me thinking: I wondered why it was that doctors who charged so much for their services, and put on such magnificient and autocratic airs, were so incapable of curing people who trusted them with their lives.

Another episode of medical importance occurred in the family about this time.

My father was the secretary of a world-wide benefit lodge. He had been a member for quite a number of years and paid quite enough to hire doctors for many months, that is, estimating on their actual value in cash, not what they usually charge if they think you can afford it.

The benefits were supposed to be available for the whole family. (And I suppose one joins a benefit society for its benefits.) We were a pretty healthy lot with whom merchants did a much better trade than the medical faculty.

However, my mother fell ill of some complaint which caused my father some anxiety. In consequence he sent, at about nine o'clock in the morning, for the doctor of the lodge.

On arriving home at five o'clock in the evening he was surprised to learn that the doctor had not put in an appearance, so he sent one of us along again, requesting his early attention. To this father received no reply. Although he waited up late, there was no sign of the doctor.

In the morning he went himself to ascertain why the doctor had not come, and was coolly informed that, because he was a mine manager, he should be in a position to pay for a private doctor.

My father told the doctor, without equivocation the opinion he was compelled to form of him, engaged a doctor in private practice, and, at the next meeting, resigned from his lodge, giving his reasons for doing so.

In justice to the lodge, I must add that the doctor was expelled from its employ, and was very angry at losing so remunerative a job that gave him little trouble.

Young as I was at this time, I was hatching an idea that the medical profession was built on a wrong basis. It appeared to me that their aim was money without service, or as little service as possible for as much money as possible.

Many events of my later life went to confirm this impression.

"The Corner" c.1864 Ballarat's Street-Market Stock Exchange, where shares were traded in the street, on the footpath and in the lane by the Unicorn Hotel.

The same "Corner" c.1882. The South-Eastern corner of Sturt and Lydiard. Trading continued in the Mining Exchange and around the Mechanics Institute (featuring the decorative facade).

4

School And Work (1875 - 1883)

When I was about twelve, my elder brothers and I were sent to Mr Menzies' High School. [1]This meant a walk of about four miles, but we did not mind this. In those days, before bicycles, trams, and motor cars were introduced, walking muscles were so developed that walking long distances was regarded as part of the day's routine.

Mr Menzies was one of those hairy monsters who had adopted as the motto of his life, "Spare the rod and spoil the child". He had a temper that was quick and violent and, when his wrath took fire, his eyes protruded from their sockets in such a manner that a practiced hand with the single-sticks could have cut them off with one stroke. And his coarse moustache would rise and stand straight out "like quills upon the fretful porcupine."

One day my eldest brother, who was about seventeen and about six feet high, studying for the Matriculation Examination, was performing some trifling misdemeanour when Menzies saw him. Out went his eyes, up went his moustache, and a stentorian voice roared.

"You think that I am afraid of you, do you? If you were as big as an ox I would not be afraid of you." And away went a heavy ruler hurled with all his force at Alex's head. The head ducked, the ruler missed, then bang, it went through the window. The quite irrepressible titter that went around the school only increased the old man's wrath and his fretful porcupine's quills remained in evidence for the rest of the afternoon. There was no peace for the rest of the session.

I fancy that the capacity for imparting knowledge is not commensurate with the name of the school viz. High, as its records at examinations were very poor. Neither of my brothers was able to pass the Matric. And they both possessed brains - as their later life proclaimed.

I did not remain long at Mr. Menzies' school. One day he got the porcupine touch with me, and gave me several cuts with a cane across the hand. They left great black weals that remained in evidence for over a fortnight. My father noticed them and, after hearing my story, gave me the choice of going to the State School which was within half a mile of home, or continuing at Menzies. Of course I chose the State School. I never regretted it.

I got on well there and, but for the aroma of moleskin pants, and the ignominy of often having to wear my father's old hat, I look back on my school days with many pleasant recollections.

This was where I got my first lesson on temperament. I was a stubborn kid; if I made up my mind to a thing I could not be shifted. For some disobedience I was awarded the punishment of writing forty lines after school hours. When the time came to start my penance I refused to write a line.

My teacher Mr Herbert Bechervaise, a very fine gentleman, at first treated the matter lightly and assured me that he was in no hurry, that he would await my pleasure. He walked up and down the room whistling and singing, and occasionally advising me to start my duty which I would have to do it before I went home, if he stayed all night with me.

A couple of hours went by thus. The sun went down. Mr. Beachervase said he would send out for candles directly and that we would have an all night sitting. But this did not move me, there was no giving in on my part. Then he came up, sat beside me and put his hand on my shoulder and said in a kind and gentle voice, "Listen to me, Willie, you are a very stubborn boy, and your stubbornness will lead you into a lot of trouble. You are punishing yourself more than anyone else, and, unless you try to conquer that temper, you will be a very unhappy boy. Now let me see you be a kind and sensible fellow and write those lines."

This was too much for me, it broke down all my defences and, with tears in my eyes, I picked up my pencil and began to write.

"That will do now," said Mr. Beachervase before I had written half a dozen words, "We will adjourn now, and I trust we will be good friends." We were. And that was the end of my stubborn temper.

I made very good progress in this school, for the system of State education was very good, even in those days of its early introduction. There was much talk among a certain class, when free and compulsory education was being inaugurated in the State, that these schools would be for the pauper class and be practically charity schools. But it was borne on the public mind that this impression was entirely erroneous.

For one thing there, was practically no pauper class extant. For another, the public was beginning to realize that the education provided by State Schools was, in a measure, superior to that provided by many of the private schools. And children of the successful business men and members of parliament were attending these schools; the stigma of Charity was being quickly eliminated.

At about fifteen I had reached the highest class in the school and obtained my certificate entitling me to leave.

Mr Nichols, the Head Master, who no doubt noticed that I had some ability for mathematics and allied subjects suggested that I should go to college and qualify as a Civil Engineer. He advised me that this would mean about seven years at College and University.

In informing my mother of this suggestion, I got very little encouragement and, of course, abandoned the idea. She then asked me if I would prefer to go to work, or continue at school.

Because I was anxious to have a good education, and just beginning to fully understand the intricacies of most subjects, and educational problems were, in consequence, becoming more interesting, I intimated my desire to continue at school.

"That's because you are lazy!" said my mother.

"Well, if you think that, I will go to work." And so I left school. My father obtained me a position as apprentice to an engineering firm named Lonie and Dingle.

My wages were 2/6 per week. For this I fired the boiler, drove the steam engine and, in the intervals, screwed bolts with what are now called common dies, the modern lightning stocks and dies not having then been introduced.

Shortly after this I crushed my finger and, during my enforced holiday, applied for and obtained a job at George Munro's Agricultural Implement Works

George Munro's Agricultural Implement Works at Alfredton (5 miles from Ballarat) c.1881.

Ballarat School of Mines   Founded in 1870 as Australia's first technical and third tertiary institution.
At the opening ceremony Sir Redmond Barry stated that Ballarat with a good School of mines, would be immortalized, and an intellectual stamp more valuable than gold would be placed upon society.

at Alfredton. This was about five miles from my home. Consequently, I lived at Alfredton and batched with my cousin and James Forbes in a small house we rented.

My wages were now 5/- per week, which was a small fortune to me.

My experiences at Munro's were rather varied and amusing. George Munro was a large, broad shouldered, level-headed Scotsman. His brother Archy was foreman of the shop.

They were quite famous as agricultural machinery manufacturers, and soon learned that they had the best machines on the market at that date, not because they had any marked ability over their competitors, but because they were never too wise to learn. They got most of the ideas from old farmers who knew what was wanted but not how to get it.

A visiting farmer said one day, "Why don't you fix something on your Winnower that will put the grain into the bag?"

Then George's brains started working, with the result that a small elevator and bag holder were fixed on the Winnowers. They thus became a machine for separating grain from chaff, and bagging the grain. This was patented as George Munro's idea. George never missed a chance like this and so made good progress.

We were very unlucky one day. Being early for work, we thought to play a joke on some of our shopmates. We fixed a can of water over the top of the door so that when the door was opened it would tip and drench our victim. Imagine our consternation when we saw the door open and the water deluge upon who? George Munro. He raved and roared for a while and wanted to know who did it. But we were all loyal so the culprit was never discovered. Being free agents as we were, away from the influence of home, we were always up to some mischief such as robbing gardens, or stealing boats from amorous lovers who would leave their boats, with the oars in, on the wharf and go to sit on some quiet and not too well illuminated seat in the gardens. I realize this was a mean thing to do, but it was only a boys' joke at that time.

The Band of Hope was our great entertainment. There was one branch in Alfredton, one at Wendouree, and another in Pleasant Street. We used to attend all these. I am afraid the chief attraction was taking the girls home through Albert Park or the Botanical Gardens after the meetings. However, they were good fun in themselves, and no doubt did a little good in the temperance cause. Evidence pointed to the fact that the excursions through the park and gardens had some semblance of Burn's Holy Fair in as much as "Many jobs began that day, to end in Hogmagandy some other day." What Hogmagandy is, must be left to the imagination.

This I know, that no less than five young fellows earning from 10/- to 15/- per week did the honorable thing and married the girls. Their weddings all happened within a few weeks of one another. Our good boss, George Munro, raised all their wages to 25/- per week, thus enabling them to make some pretense of keeping the legendary wolf from their doors.

These episodes gave the rest of us a scare and warned us to be careful. We had so far escaped, but I must say there were a few girls among our Band of Hopers, who, to use the words of the Sentimental Bloke, "Chucks her carcass at a feller's head, and mauls 'im." One of these, a fine buxom lassie she was too, several years older than I, tried very hard to lead me on. But the spectre of those five premature marriages, together with an ambition of my own, made me afraid, and thus saved me from a similar disaster.

These were happy days, jolly days, we made the most of what little fun there was. And one, at least of the friendships made then lasted through life.

This was with Jack Dobson, a fine big athletic boy, with a sterling character. He afterwards filled his home with trophies won by his great prowess at rowing, but he could not be induced to study, although I was attending twice a week evening classes of the School of Design. The way Jack would enter the middle of a ring, stop a fierce fight between two irate kids then make them shake hands and go away together as friends used to evoke my undying admiration, and made a lasting impression on me.

What impressed me most was the fact that older men were in the ring enjoying the fun and encouraging the kids to greater energy, when a boy of seventeen or eighteen enters as The Good Samaritan, and shames the lot of them.

Another circumstance during my stay at Munro's impressed me very much. All the agricultural machines were painted in glaring colours. Red and blue predominated, show machines were richly decorated with flowers and elaborate scrolls. This was not in accordance with my taste in art, but that fact did not seem to have any bearing on the conditions. I suppose these gaudy colours pleased the average farmer, and that was, of course, essential.

There were two painters who seemed to be quite good men, and very honest diligent workers. But, no doubt the paint was to some extent a luxury, and George Munro a little impatient, so he rounded up the painters one day and told them they were not turning out enough work. They took the matter philosophically and suggested he let them work on contract - they would show him what they could do.

These men's pay would be about £3 per week. A contract was arranged which, on the cost of the work previously done, was in Munro's favour. Then these men got going. They used a brush in each hand, and the paint supply rolled on to the machines.

The first week's reckoning disclosed the fact that they had earned something like £22 each. This was such a shock to George that in his anger he cancelled their contract, and fired them off the premises.

I stayed only about eighteen months at Munro's. Then my father obtained a job for me at the Phoenix Foundry as an improver. This was at that time a locomotive works. My first pay was 10/- per week.

This was a fine shop doing good work and maintaining a firm discipline. However, I had not been there a week before Joe Wells, a bolt maker who worked a steam oliver, came over to me and called me a "new broom". I asked him what he meant. He said that I was working pretty hard, but he guessed I would soon get tired of this and settle down. I took no notice, but afterwards found that the policy of most men was a fear of doing too much work. I never developed this fear.

There were quite a lot of real good fellows in this shop, and it was here that I began to associate with real good companions.

My foreman was a Scotsman named Hughie Reid. In fact, all my bosses in these shops were Scotsmen. I remember that we attributed this circumstance to a Scottish trait that they did not mind the pay so much as long as they had a "wee bit authority". Hughie was an excitable bustling chap who very seldom came near any one without saying "Hurry up now, lad, I want this engine out on

The School of Mines, Ballaarat.

Ingenio effodere opes.

This is to Certify

That William George Manners of Ballaarat Victoria —— whose signature is hereto appended, has passed the Examinations of THE SCHOOL OF MINES, BALLAARAT. and that the Council, on the report of the Examiners have awarded to the said William George Manners in accordance with Article 4 — of Statute passed 24th day of October 1872 this Certificate of competency as an Engineer. ————

Signature WG Manners

Given under the Seal of The School of Mines, Ballaarat. this Eleventh *day of* January —— 1887.

Andw Berry
REGISTRAR

William F Stawell
PRESIDENT

Saturday." We were turning out one locomotive engine a week. He "hurried up" Dave Carnegie one day, and fell in badly.

Dave was a dramatic artist who undoubtedly had great ability in this line. He was the leading light in the Ballarat Amateur Dramatic Club, having played Rob Roy, Pizzara, and others with marked success.

Dave's tragic demeanour was characterised in his gait and actions. He walked with a big springy stride, swung his arms dramatically, and held his head with dramatic dignity. He was equally good at his work, and had been working on eccentric straps and sheaves for a long time. He perhaps could do as much on this job, owing to his lengthened practice, as any other two men in the shop.

But Hughie couldn't help it, it had become a habit with him. So, passing Dave one day it came out. "Now hurry up, Dave, my boy. You're no' doing half enough, you know. I want this engine out on Saturday."

Dave stopped, straightened up, and glared at Hughie for about five second swith his best stage glare. Then with a graceful sweep of his arm, reached his coat down from the wall and, in the Hero's voice, and excellent emphasis, exclaimed. "If I can't do sufficient for you, get someone who can." and started to stride with the Villain's exit out of the shop.

Hughie immediately collapsed and trotted after him, saying, "Here man. Davie, I'm no' meanin' it. Come back to your work, lad, come back to your work. I'm no' meanin' it ata'." Then Dave stopped, glared at him again with a look that made Hughie squirm, returned with the same dramatic gait to his bench, hung up his coat with the same graceful sweep of his arm, and started his work, while Hughie sneaked off and no doubt felt a worm. It is not on record that he ever reprimanded Davie again.

The Company to encourage their boys to attend the School of Mines paid two thirds of the fees, viz. 14/- per quarter. The student paid the 7/- balance. So I took advantage of this to learn Mechanical Engineering and Draughtmanship.

Hughie Reid was our teacher. I worked under him for five years before I discovered that he was not teaching us anything except to copy mechanical drawings.

At this time I had to walk two and a half miles to work, and the same distance back in the evening to the School of Mines, twice a week. This meant ten miles on those days in addition to 8 hours pretty hard work. Now, when I think of having done this for 4 years, to learn so little, I look back on those days with regret.

One of the boy's fathers evidently discovered this anomaly, and complained to the Board, with the result that Hughie was instructed to give lectures. Of course by "lectures" was meant lessons on Physics. Natural Philosophy, Strength of Materials, Laws of Gravity and Motion, Statics, Dynamics, and allied subjects. But I fancy these were a closed book to Hughie. He was a fitter only, had no other experience than locomotives, and knew nothing else. His only two lectures are worthy of record.

Our class worked from 7 p.m. to 9 p.m. and about 8.30 Hughie said, "Now lads put your tools away, I'm going to give you a lectur'."

When all was ready he hung the plan of a locomotive on the black board and started. "Now, lads, I'm going to give you a lectur'. This is a locomotive," pointing with a cane, "This is a biler, this is the funnel, this is the engine cylinder, these are the brakes," and so on, right through the machine, telling us the name of each part. We, who had been some years in the loco shop were, of course, not greatly enlightened. The burden of knowledge which he had imparted did not cause any headaches.

Hughie's next "lectur" was a similar fiasco. On this occasion he hung the plan of a loco tender on the board and started, "Now lads, I'm going to give you a lectur. This is a tender, this is the tank, this is the sand box, these are the buffers, these are the wheels," and so on right through.

Whether the Gods of Wisdom were looking down or not the School Board was looking up and Hughie got the sack. After this, Mr Summers was appointed and we began at the beginning again, learning the rudiments of Natural Philosophy, Geometry, and Steam and the Steam Engine, advancing later to the higher studies.

Thus I had spent seven years at the School and had to crowd into the last two years as much knowledge as possible. However, I got my diploma as a Mechanical Engineer, and Draughtsman.

I was always a great favourite of Hughie's. This was because I was his last pupil as a draughtsman, and a diligent worker in the shop.

Some of the "Hurry up" periods recurred to me, however. I was working under a fitter named Frank Davies. Frank was head fitter on the tenders and was one of those idiots who rush at everything and lose more time, through mistakes, than would do the job if brains were used more than feet. Hughie would come up and say, "Hurry up Frank, I want this tender out on Saturday." As a matter of fact, Tuesday was the day for sending out engines, but Hughie always wanted them on Saturday. On being thus admonished Frank would rush and tear around, marking out work for the machines in such a hurry that, when it came back from the machine, we would have to spend hours and perhaps days chipping off what should have been set out for the machine. Sometimes, even, the work was so far wrong that it had to be sent back to the machine.

Frank was always late and always in trouble and under the "Hurry up" injunction. I was then in my last year, and had been on the tenders for some time. Hughie shifted Frank and put me in charge. I was very nervous, at first, but Hughie accidentally cured me of that and gave me a lesson that served me to good purpose all through my life.

I had a habit of asking Hughie every time he came up, how to do something. It was, "How would you do this?" or "How would you do that?" Reid always very patiently told me, until one day he was evidently in a bad temper when he arrived, and, of course, I had some silly question to ask. He snapped out, "Think for yourself, lad, think for yourself. What's your brains given to you for? Think for yourself." and he dashed away.

Thanks, Hughie, old man, thanks for one of the best lessons I ever received. I thought for myself in future, never asked any more questions of anyone but myself, and found that by thinking for myself I could solve all the problems of life, and find the solution of many questions which otherwise would have escaped me. It is the man who thinks for himself who gets on in the world. And, although I have not done anything great, that one principle has lifted me above the common rut.

Hughie would come to me and say, "Now hurry up, Bill, old man, I want this tender out on Saturday."

"Yes Sir." I would reply, and continue in the same diligent methodical way, and have his tender ready for him on Monday evening ready to take its place behind the loco on Tuesday, always, without fail.

When we were sending out the hundredth locomotive the shop was stirred into a gala show of unbounded enthusiasm. It was a red letter day in the history of Ballarat; several Ministers of the Crown, including the Premier of the State, Mr Alfred Deakin, Mr Duncan Gillies and Mr Graham Berry were invited, also several leading citizens, including the Mayors of Ballarat City, and Ballarat East.

Because the shop was some distance from the railway station, rails were laid in the street and because there was not sufficient room to put in a curve after leaving the shop door, a portable turntable had been constructed. On occasions of sending out engines, this was erected in the middle of the road, the engine run on to it, turned at right angles and run down the street with men carrying red flags warning the public of its approach.

On this momentous occasion Hughie Reid was to have that honour. The loco was richly decorated with flowers, bunting and patriotic legends. I had the distinction of painting two banners which stretched from the funnel to the cab, issuing the command ADVANCE AUSTRALIA. There is no doubt it was very artistically and beautifully decorated.

The streets were crowded with sight-seers awaiting the great event. All the distinguished visitors took their places on the foot plate of the locomotive with W.H. Shaw, the General Manager, a bluff, stout, old gentleman. Steam was up and all pronounced ready. Of course, the engine had been under steam the previous day and had been well oiled and greased. And, of course, condensed water had lodged in the cylinders and pipes.

It is usual, as any engineer knows, when starting a steam engine from cold, to open the release cocks and blow all the water out of the cylinders before putting on sufficient pressure to start the engine.

But Hughie was excited as usual. Was he not the great hero of the day, the cynosure of all eyes? He forgot to open the release cocks. The big front door was open, the whistle sounded, and a cheer broke forth among the crowd. Hughie turned on the steam. Puff! Plop! Plupp! roared the engine. A magnificent spray of dirty water and grease issued from the funnel, and descended in a devastating cascade over the surrounding top hats and frock coats.

The scatter that was made, jumping from the foot plate, put up records for agility among some of the fat politicians, and must have strained many gouty limbs.

Mr Shaw's remarks cannot be written down for public absorption. If it were a court case they could be written and handed to the magistrate.

It was quite a while, after the release cocks had been blowing for some time, that the visitors could be induced back on to the foot plate, and the engine duly driven on to the turntable.

Dave Carnegie was selected to make the speech on behalf of the men and, after the Premier, Treasurer, and Mayors had spoken for the country and town, and Mr Shaw had spoken for the management, Dave came out in dirty blue overalls with unwashed hands and face, a dirty piece of cotton waste in his hands, and made the most eloquent, interesting, and oratorical speech of the day, quite putting the great Alfred Deakin in the shade. But then, you see, Dave had not had his ardour damped by Hughie's impromptu shower bath. And Dave was, in any case, a born orator and elocutionist.

I had pictured, at this time, Dave as a Premier of the State, and a great career as a Statesman. But this was not to be: Dave arrived one morning at work, having been converted to Religion the night before, and went round trying to convert everyone in the shop. He gave up all worldly affairs, including his union and life insurance policy, and became a preacher.

He died quite a young man, of gall stones. Dave would not consult a doctor because he was in the Lord's hands. He would lie in bed suffering the excruciating pain of passing his gall stones, and cry, "Pile it on. Oh, Lord, punish me, persecute me, I have been a sinner and deserve it. Pile it on, Lord, pile it on."

The Lord evidently took him at his word for he piled it on so thick one time that it was beyond Dave's endurance. So he passed away praising God. One of the finest, ablest, and most honestly sincere men, I believe that it has ever been my pleasure to love and honor, became a memory.

Another "Hurry Up" affair of Hughie's may be worth recording as showing the futility of rush and tear, against calm judgement and quick administration.

The tenders were carried on three pairs of wheels, the axle boxes were carefully fitted to the axles, and the "rubbing pieces" as we called the cast iron plates between which the axle boxes worked, were carefully fitted to the frame, and then the whole machine was lifted by an overhead crane at each end then carefully lowered on to the axle boxes. I had six men and boys on this job, one to manipulate each axle box. They were a typical lot: There was Joe Gates, a farmer's son who was a local preacher in the Wesleyan Church. He had a big voice and wore a beard and looked a typical hayseed. We nicknamed him "Farmer," and "Whiskers", mostly the latter, he evidently did not fully appreciate our compliments on this tonsorial beauty, and arrived one morning with a clean shave. Holy Smoke! we now realized the necessity for the beard. He had practically no chin, a "caricature of a face." So we now christened him "Chin", and his beard grew again.

Then there was Harry Haigh, another local preacher about seventeen. He was a fine looking boy, who came to work in a padjet coat, hard hat, and white shirt and collar. We called him "The Toff". He once said to me, "How I wish I was rich and had plenty of money."

"What for, Harry?" I said. "You preach that a rich man cannot enter the Kingdom of Heaven."

"Oh!", he replied, "I would like to be rich so that I could give to the poor."

"Why not wish the poor rich, so that they would not require your degrading charity? But then Harry, you would not have the proud dignity of patronising the poor, would you, Harry?"

Then there was Alf Bryant, another boy who loved me like a brother. He was a fine lad full of good principles and a kind heart. My chief recollection of him is when I left the shop for good, tears rolled down his cheeks as he bade me good-bye.

The other three were nicknamed Long Dick, Fat Dick, and Little Dick. Long Dick was a man called Warland. He was about six foot three inches in height, and proportionately strong. He was a "bit of a lad", afraid of no man, and was always chewing tobacco. Fat Dick was a boy called Davies, a fine boy, a good singer, but resembling the fat boy in Pickwick in his proportions. Little Dick was a little old Cockney, a miserable specimen of humanity who was supposed to be a humourist. He sang songs and gagged at smoke socials. His items were most appreciated by the fastidious Sunday School teachers.

Well, to get on with the tender, each man was at his box, and the frame had been lowered until the rubbing pieces were an eighth of an inch above the boxes. We had begun to adjust the wheels so the boxes would be directly and absolutely under the rubbing pieces, then they could be lowered into place without jamming.

When Hughie came on the scene, "Hurry up now, Bill, I want this tender out on Saturday." Then, to the crane men, "No.1 crane, lower! Lower! Lower! Hold! Hold! Hold! Hold!" the boxes were jammed. "Heave up! Heave up! Hold! Hold! No.2 crane, lower! Lower! Hold! Hold! Hold!" back end jammed. "Heave up! Heave up!" so he went on roaring "Heave up! Hold! Lower!" until all the boys were excited, lost their heads and, of course, made no progress. Then Long Dick rose from his box, spat a blob of tobacco juice over his shoulder, then said in his careless drawl.

"I say, Hughie, give them time to smell."

Hughie glared at him a moment, turned on his heels, and rushed down the shop. Then we quietly got things organised again. "Lower gently No.1 - hold!" boxes nicely entered. Then, "Lower gently No.2 - hold!" boxes nicely entered. "Lower gently altogether." In five minutes the job was finished.

1. Next to The Western Fire Brigade building, N.W. Corner Sturt & Raglan Street.

FINIS PAR EXCELLENCE

OF THE

Smeaton Agricultural Show,

ON FRIDAY, 12th NOVEMBER, 1886.

GRAND

CONCERT & BALL !

IN AID OF

Burns' Statue Fund,

IN HONOR OF

SCOTLAND'S GREATEST PLOUGH-BOY POET,

WILL BE HELD IN

Anderson Bros.' Barn, Smeaton,

AT EIGHT P.M.

The Artistes for this Concert will comprise the **principal exponents of Scottish song that Ballarat can produce,** with

Mr. Wm. KING as Conductor,

Whose name alone is sufficient to justify the most sanguine expectations; whilst during the evening the Entertainment will be interspersed with specimens of

ENGLISH MELODY

ELOCUTIONARY ART

And several gems (both comic and sentimental) of

IRISH MINSTRELSY !

THE BALL

Which follows will be conducted on the strictest lines of PROPRIETY. First-class Music and Professional M.C. provided. **Refreshments Supplied during the Evening.**

PRICES OF ADMISSION.

CONCERT: 2s., Children Half Price. BALL: Single Ticket, 3s.; Dcuble Ticket, 5s.

TICKETS may be obtained anywhere on Smeaton; Mr. Geo. Tait, Storekeeper, Creswick; Mr. R. Rail, State School, Dean; Mr. D. McGrath, Storekeeper, Allendale; Mr. John Eales, Storekeeper, Kingston; and Mr. Wilkins, Storekeeper, Newlyn.

F. W. NIVEN AND Co., Printers and Lithographers, 34 Sturt Street, Ballarat.

5

Quacks And Politics (1880 - 1885)

While at the Phoenix Foundry when I was about eighteen, a very distressing affair occurred to me. I recorded it on the grounds that it bears a subject which hides its head from public view, and so a menace to the public weal goes on unchecked from generation to generation.

Victims of its nefarious practices, when caught in its meshes, suffer its penalties and ignominy in silence for very shame. And so vipers who prey on the ignorance and credulity of unsuspecting youth flourish and prosper without let or hindrance from ruling authority or public opposition.

I was a strong healthy lad, as one might assume when I was eleven stone at eighteen, and could lift 100lbs. in one hand straight above my head. But, no doubt owing to the mental distortion of the influence of my earlier companions I was subject to rather frequent sexual dreams and their result. As my health and strength indicated, the effect was by no means serious.

But the papers were full of nervous debility advertisements of medical specialists and quack medicines. They published lists of symptoms of this allegedly dire disease, which covered all the little ailments that flesh is heir to: spots before the eyes, headaches, dizziness, tired feeling, and a hundred others, most of which experience tells me are the result of indigestion, colds, and other small but prevalent ailments.

Added to these, discussions with young wise fools made me feel that I should see a doctor. Taking advantage of a picnic to Melbourne, and having saved about £3, I consulted a Dr. La Mert. He examined me, admitted there was nothing wrong with me, but by strength of will and domineering put words into my mouth and forced admissions from me which he wrote in a book. I suppose this was for his own protection. He gave me a bottle of medicine, charged me £2/0/0, and advised me that, on sending a guinea, I could get a further supply.

I returned to Ballarat, took his medicine, and sent the guinea for another bottle. By the time I had swallowed this I felt I had lost some of my vital energy, but the dream effects were not lessened to any extent.

I had another guinea, so I bought a bottle of Dr Ricords Essence of Life, and swallowed that. This put the lid on, as the saying is. I found, that I would be totally incapacitated from consummating my conjugal obligation. My vitality from a sexual standpoint was utterly destroyed. But unfortunately the dream discharges went on all the same. My distress was appalling, and although I was very fond of the lassies in a moral sense, I could see a life of celibacy in front of me. So great was my distress that my elder brother, who slept in the same bed with me, found me sobbing when he came in. By persuasion he induced me to tell him the whole miserable story.

How he raved and lectured, calling me all the kinds of fools that ever tried to live. Told me that those wretched debility specialists were parasites and rogues who traded and made fortunes out of such fools as I. "Why didn't you come to me? You fool! I could have helped you, and saved you from these despicable cads."

The next day he bought me a book by Dr Cowen on the *Science of a New Life*, which told me all about myself from a natural, religious, and scientific point of view.

I thus learned, at eighteen, that which I should have been taught by my parents at fourteen or fifteen, and thus have been saved this very distressing, and heartbreaking experience. By following the instructions of this book, by a clean employment of the mind, and cold baths, I ultimately restored my health and found that in all these things nature is wonderfully recuperative.

At the same time many may not have a brother like I had, or a friend, or parent to come to their aid. Their disaster is life-long, and leads to unhappy homes or miserable old bachelorhood.

I contend that all these parasites should be debarred by law from trading on public ignorance. To abolish ignorance would be better, but it is difficult. The wise should try to remove snares and traps from the paths of the ignorant, and thus save society from a permanent and aggravating menace.

Just one word to parents before I go on to politics, and that is, teach your children as soon as they arrive at the age of puberty, the beautiful and all-wise provisions of nature for the reproduction and continuance of the race. It is better that they learn it from you than from the highways and byways, brothels, and lewd companions. If you feel any diffidence in broaching this subject to your children, and I admit this too, is difficult, send to the White Cross Society of Sydney. They will send you suitable literature which will serve the purpose in a real and effective way.

The Victorian legislature had recently introduced a measure giving the people Manhood Suffrage and, because I was approaching the interesting age of Manhood, I would be entitled to a vote. But, with or without a vote, electioneering in the Ballarat East constituency included that interesting bit of Ireland (planted overseas) named Bungaree. This farming district contained a population of the wildest Irishmen I have ever met. They walked with a stride, carried shillalahs, and glared at anyone who looked like a Protestant, hoping for the slightest provocation to crack his skull. They, added to the Roman Catholics in Ballarat East, made certain that the contest for the seat between the Irish and the Protestant candidates was close and fierce.

Mr C.E. Jones, who was a great Wit, had christened them Bungaree Savages some time previous to my entering the fray. He was a candidate for the Ballarat East seat and was advertised to address his constituents in the Town Hall on a certain evening. A mob of Bungaree's best Fenians came in to create a disturbance and prevent Jones speaking. This is always the tactics of folk who have a poor case And I notice that it is often adopted by the Communists against their opponents, and the liquor trade, against the temperance forces.

It was a big meeting. The Chairman briefly introduced the Candidate. C.E. Jones rose and said, "Gentlemen of Ballarat East."

"Boo-oo-oo." roared Bungaree, and kept it up for some time.

Jones calmly waited for a lull then began again, "Gentlemen of Ballarat East."

"Boo-oo-oo." again from the Fenians, then a second lull. "Gentlemen of Ballarat East."

"Boo-oo-oo." with special vehemence.

At the next pause Mr Jones added, "And savages of Bungaree." henceforth they were called by this name.

My first experience with them was when Mr Dan Brophy fought Mr John James for the seat. The contest was fierce on both sides and, on election day, the "Savages" assembled in great force. Towards the close of the poll, beer and poteen were having their effects; the Phoenix bar was crowded with their drinking, and arguing. We boys had got in to watch the fun. One man came in and dashed through the crowd waving his shillalah round his head, and roaring "I'm a Fenian, heart and soul I am".

But this did not stir them to any extent. A little later a wild eyed man rushed in and bawled, "We've been defated." and immediately hit the nearest man on the head with his Waddy. Like a flash everybody lashed out. Pandemonium reigned. It was a case of where there was a head, hit it. The noise was terrific. We got behind the bar, but not before our clothes were bespattered with blood. Then several police came in and, with their batons and handcuffs, slashed about them making matters ten times worse because the "Savages" had someone else to fight now. And so the ghastly brawl went on with increasing vigour for some time.

Then Dan Brophy came and climbed on to the bar, and put out his arms.

"Byes! Byes! Byes!"

"Hooroo! for Dan." then burst forth.

"Byes," says Dan, "silence now." and all was calm. "Byes, we have been defated."

Chorus, "Hooroo for Dan."

"Byes." said Dan, motioning with his hands for attention, "Byes we have been defated, but let them see that we can take our defate like min. I want you now to all go home to your wives and families, the same as I am going to do, and we will come back next iliction and show thim phat we can do."

"Hooroo for Dan, Hooroo!"

"Silence now, byes, go home, go on, byes." And they marched out like a flock of sheep.

The power of a strong man over a mob is something marvellous, but the power of an Irish leader over the ignorant rabble is a miracle.

Dan Brophy was a pretty good man although it is on record that he could neither read nor write. He was a Publican. It was a strange fact that nearly all the publicans and practically all the police were Irishmen.

A good story is told by Mrs Brophy and, although it is years old, may be worth recording. She was anxious that her daughter should pass the Civil Service Examination, so sent her to College for that purpose. Her injunctions to the Principal were very explicit and forceful.

"Me darter must pass this exam." she said. "No expinse must be spared. Buy her phativer she wants. Don't wait to ask, buy it and sind in the bill. She must pass widout fail."

When the time came the daughter failed, and Mrs Brophy was, to put it mildly, annoyed. She blamed the teacher, of course, and called at the school to visit her justified wrath upon him.

"And whoi the divil didn't she pass?" Mrs Brophy asked in an indignant tone.

"I'm afraid," replied the professor, "she did not have the ability."

"Thin whoi the divil didn't ye buy it fer her? Didn't I tell ye to spare no expinse?"

But of all the natural born humourists, Ned Murphy and his wife were supreme. Dan Brophy did not offer his services again, and Ned Murphy stepped into the breach. Ned, too, was a publican, which statement might be superflous, but he had risen to that noble and inspiring profession from the ranks of Navvydom. His rise was recorded by Mrs Murphy and she was justly proud of her Ned.

She said, "When I met Nid furst he was plain Nid Murphy a railway man. Then he got into a pub and he was Mr Murphy. Thin he got in the Council, and was Mr Alderman Murphy, and now he's in Parliament and is Mr Parliamentary Murphy, so phat relation am oi to the quane?"

Elections came round again. I had a vote this time, so attended the meetings. The Savages were as much in evidence as ever. Mr James was the Sitting Member, and was having a hard fight for it. The Savages attended all his meetings in force; there was no peace.

Mr Murphy's meetings were large and enthusiastic and there were no interruptions. Discretion is regarded as the better part of valour, and the man would have indeed been indiscreet to interrupt Ned, with a Fenian nursing a huge shillalah, sitting along side of him. He could have only done it once, and then the ambulance. Then Ned could express his sentiments freely. Of course his speech was written out for him by the priests so it was all right, and free from humour. It was at question time that Ned shone as a humourist.

"Do you believe in the introduction of the Bible into State Schools?" asked one.

The Murphy, "No, be jabers! the Bible is only for bigoted, narrow-minded Protestants."

"Do you favor water conservation?" asked another.

"Water conversation? No I don't. What's the good of talking about it? Let's act." replied Ned.

At this time the great Mildura Irrigation Scheme was being formulated and, of course, much discussed. But perhaps Ned did not read the papers.

"Do you believe in irrigation?" asked another.

This was a question Ned was emphatic on. He rose, thumped the table, and replied. "No I don't. There's plenty poor people in the country already, widout bringing anny more in."

Ned was elected, and Ballarat East was, of course, not represented in Parliament, but Bungaree was. Later at a re-distribution of seats, Bungaree was cut out of Ballarat East. Thus this latter constituency became civilized and uninteresting. John James became a Minister of Education, and filled the Schools with Cornish teachers. Ned became the permanent member for Bungaree.

ANNUAL DENNER

O' THE

Scottish Sheep's-Head Club,

HELD IN THE

AULD REEKIE HOTEL

ON

AULD HANSEL MONDAY, 1886.

THE HONOURABLE LORD SHEEPSHANKS
in the chair.
BAILLIE SOURDOOK, Croupier.

BILL OF FARE.

SOUPS.

Sheep's Head Broth. Sheep's Head Kail.
Sheep's Head Tattie Soup.
Sheep's Head Turtle. Sheep's Head Cockie-leekie.
Pudden Bree.
Cauld Kail *a l'Aberdeen.*

FISH.

Kipper Salmon. Kippered Herrin.'
Saut Herrin.' Labsters. Buckies. Limpets.
Cockles. Finnan Haddies. Caller Haddies.
Caller Ou.' Caller Herrin.' Red Herrin.' Garvies.
Speldrins.

ENTREES.

Sheep's Head Parritch.
Oatmeal Parritch (with milk).
Oatmeal Parritch (with whey).
Cauld Parritch (with Glenlivet). Pease Brose.
Oatmeal Brose. Athol Brose. Kail Brose. Sowens.
Trumlin' Jamie.

REMOVES.

Haggis. Trotters. Tripe. Collops. Pork Steaks.
Sheep's Puddens. Sheep's Bags. Sheep's Shanks.
Lichts. Harrigles.
Stoved Tatties and Soo's Ribs. Cauld Tatties.
Black Puddens. White Puddens. Sma' Ham.

DESSERT.

Tattie Peelings. Tattie Scons.
Nicket Baps. Soda Scons. Farls o' Cake.
Crackers. Cookies.
Snaps. Sweeties. Gundy. Treacle Candy.
Dog Hips. Cat Hips.
Haws. Black Man. Sugar Bools.
Carvies. Peppermint Draps.
Jellie Pieces.

CHEESE.

Skim Milk. Gowdie. Daighy. Mitey.

WINES.

Whisky Hot. Whisky Cold. Whisky New. Whisky Old.
Whisky Neat. Sugar-Auly Water. Shandy Gaff.

WHISKY TODDY.

Tallisker. Treacle Yill. Skeichan. Soor Dook.
Gregory's Mixture. (favourite blend).

6

Mutual Admiration Societies (1885)

Mutual Improvement and Kindred Societies had become very prevalent and very popular in Ballarat. There were a few independent and unsectarian, the chief of these being the South Street Debating Society. But the great majority were connected with churches. The former had a membership of men only and their debates were mostly of a political, social, or scientific nature.

But all the Associations allied to churches had a mixed membership and, consequently, their speeches were mostly made to please the ladies and elicit their appreciation.

The subjects, in the majority of cases, were founded on either Love, Courtship, or Marriage, and, although they enjoyed many years of success, mostly developed into Mutual Admiration Societies.

I was a member of St. John's Presbyterian Mutual Improvement Association, of which the Rev. J.W. Inglis was President. It was his eloquence, ready wit, great fund of humour, and love for young people that made the Society the success it was. He was charming in Scottish readings, and sketches from Dickens, and seemed to have a ready story apropos of all circumstances. I fancy he must have invented them, they were always so apt and suitable to the case in point.

These societies formed themselves into a Mutual Improvement Association Union, and competitions in music, elocution, and literature were held with great succes severy year.

This was the start of the great South Street Competitions that are now famous throughout Australia and still flourish after all these years.

Most of us who had any ability at all in that line took lessons in elocution, and so rose above the criticism that my first attempt evoked.

I won several prizes both in dialogue and recitation. I was a competitor in a dialogue on one occasion, the subject being the quarrel scene between Sir Peter and Lady Teazle.

I think I was placed, but my younger brother, who could impersonate an old man better than I, was the winner.

"Even today in Melbourne, Broken Hill, Charters Towers, Kalgoorlie, and many other centres large and small - Perth, for instance, at a farther migratory step from the Victorian goldfields - a surprisingly large number of people can claim ancestors from Ballarat."

Weston Arthur Bate 1978
"Lucky City" Ballarat 1851-1901

Note: Appendix I contains a collection of W.G.M. poems from this era, together with notes relating to his courtship experiences.

7

Broken Hill (1886 - 1887)

However, these days came to an end. The Phoenix Foundry had finished a contract and, not having another to go on with, all the journeymen were dismissed. I went to Melbourne to try my luck.

The policy of protection had established a lot of workshops before the country was ready for them, and work was scarce. I had my first and only experience of tramping from shop to shop looking for work. I had three months' work out of six, and found that my funds were dwindling away.

Among the happy days in Melbourne were a few evenings I spent with a family called Robinson in Prahran. They were real Scots; Willie Robinson would come up and say "Can ye come oot th' nicht, Wully, and hae a wee bit diversion?"

Of course I could. He had a mother and two sisters, and there would be a few Scottish friends. They would get in a supply of beer and, between strathspeys and the highland schottisches and real good rollicking fun, in which the old mother threw herself with real Scottish vigour, the time would go until the trams stopped and I would have to walk to Melbourne.

At this time the cable trams were being constructed. To form the cave under the road in which the cable was to run, elliptical yokes were formed of heavy rails, and the ends sawn to form a groove for the tram gripper to work in. These were bent by hand round a cast iron mould. They required quite a number of men to bend them, and two heats. They had a pattern maker in the shop who was very fond of beer, but who had good brains when they were not fuddled. He invented a machine that worked in an ellipse, and bent these round in a few seconds with one heat. The firm gave him fifty pounds for his invention, with which he went on a howling spree which lasted as long as the £50, that is, a few weeks. The firm made a fortune out of his machine.

There was some vague talk about Silverton, the great silver mines on the Barrier Ranges, and, I felt, a strong incentive to go there. The feeling became irresistible, and I began to make enquiries as to where it was. But no one seemed to know. One day I met Major Smith, who was the Member for Ballarat City, and mustered enough pluck to bail him up in the street. He knew my father, and gave me a very cordial reception. But when I put my question, "How do I get to Silverton"? he replied

"Now, boy, I'm damned if I know. I've heard about it, but don't know where it is. But Billy Jones will know. Go and see Bill Jones, and tell him I sent you."

Billy Jones was Mr W. Jones, the principal sharebroker in Melbourne, and I went to him with the same request. But all he knew was that you had to go via Adelaide.

"Go to Adelaide," he said, "and they will tell you".

I had a friend, George Park, who agreed to go with me, but he backed out at the last minute.

There was an exhibition on at Adelaide with cheap excursions from all parts.

So I went to Ballarat to say good-bye to my people then bought an Excursion ticket to Adelaide for 25/-. I saw the exhibition, and intended to stay a week there.

After a few days I received a wire from George Park asking me to wait for him. I replied that I would. But the next day I felt the "influence" very strongly, again urging me to go on. So, seeing an advertisement by a person wishing to exchange a Cockburn ticket for a Melbourne one, I exchanged mine, paying the extra fare of 10/- from Ballarat to Melbourne, thus obtaining a ticket right through from Ballarat to Cockburn, about 700 miles, for 35/-.

Cockburn was the end of the South Australian railway on the N.S.W. western border. The Silverton Tramway was not finished at this time, so I started for Broken Hill next morning, leaving a note at the Coffee Palace for George Park, having learned that my destination was Broken Hill, not Silverton. Broken Hill had only recently been discovered.

The journey on the train to Cockburn was the slowest thing on record. At times we could jump off, and gather flowers, and run along and jump on again. But we ultimately got there, arriving about midnight.

I had met a big Irishman named Logan on the train, who had been to Broken Hill before. We had chummed up, and I thought it wise to leave myself in his hands. Cobb & Co. were running two coaches that did the journey, thirty two miles, in about six hours.

But just as we were getting our luggage together, a big fat Irish boy came up and said "Are yez goin' to Broken Hill?"

"Yes," we replied.

"Thin come along wid me. I've got a foine pair of sphankin ponies that will take yez there in about two hours."

Logan thought it was good enough, so I agreed. We went to the pub for a drink, and saw the two coaches depart, each with four good horses. Our driver had some business to detain him. As a matter of fact, he wanted to get the coaches away before he started.

However, we were soon on the road, and before we had gone five miles the 'sphankin' ponies proved to be a pair of old crocks, who had evidently been rescued from the bone mill. They were knocked up, and quite incapable of doing more than two miles an hour.

This discovery led to argument with the result that Logan and the fat driver came to blows. It was quite thrilling while it lasted, but the fat boy soon cried "Enough!" and laid down, refusing to rise and take some more. Logan and I then walked on, soon leaving the sphankin ponies miles behind.

We arrived at the Pinnacles, about half way, at 6 a.m., and, being hungry, tried to rouse the sleeping publican, without result, and resumed our trudge. We still had fifteen or sixteen miles in front of us. By 10 o'clock we had entirely lost sight of the coach. It must have been five miles behind. About 11 o'clock a man came along with a sulky, and gave us a lift, so that we arrived in town about 12 o'clock and soon sat down to a good dinner. The sphankin ponies dashed into town about 4 o'clock in the afternoon putting up the record of two miles per hour.

Broken Hill was one street at that time, mostly consisting of temporary buildings of galvanised corrugated iron, wood and hessian, the hessian predominating. The street was paved in dust.

There was one brick hotel, the Denver City, run by Jack Elliott, who was fast making a fortune. There was no accommodation available, plenty of boarding houses, but all lodgings were occupied. By the good favour of a boarding house keeper, I was permitted to sleep under a table with a few sacks for bedding.

George Park arrived a few days later. We bought a piece of land and erected a one-roomed camp.

At this time, Mr S.R. Wilson was General Manager of the Broken Hill Proprietary Mine, and I approached him for a job, showing him my School of Mines Diplomas in engineering and draughtsmanship. He was not much impressed, didn't want anyone in that line. Then I told him I was a Fitter, producing my reference from the Phoenix Foundry.

"Ah! that's better." said he, and wrote a note to George Munro (not our farmer friend, but another Scotsman), the Engineer to put me on. I started the same day, erecting blowers for the smelters. I had not been going long before a man came up and said, "Here, matey! you'll have to pull up, or you'll work yourself out of a job."

"Well, I suppose I can get another." I replied.

Here was the same old stunt, the nucleus of the Go Slow Policy. I learned that carpenters were planing wood to shavings just to put in time. But this was no good to me. I did a fair thing and learned that I had done as much in a fortnight as my predecessor had done in six weeks.

I had not been many days on the B.H.P. when Mr Patten came out from The States to take the management of the Mine, and Sam Wilson was transferred to Block 14 S.M. Co.

I had finished the blowers after which, I caulked a long six inch steampipe. Then being finished, I asked the boss for another job. He scratched his head, and said, "You've finished that have you? Well, you know when we put steam in that pipe, we don't want any leaks. I think you had better make sure and go over it all again."

I said to myself "By jove! I've worked myself out of a job." but as I was getting my tools out to go over the job again, I got a message that Mr Wilson wanted me at Block 14. Then I found that I had worked myself into a job: I was picked from all the bunches of boss dodgers, and put in as Engineer of Block 14 at twenty three years of age.

I found then, and my experience all through life has proved, that the go-slow-dodge-the-boss policy, so prevalent among a certain large class, is not only ruinous to the workman, but spells disaster to all enterprise, and blocks progress in all industries.

I worked a few months as Foreman when Sam Wilson came up one day and said, "You can draw, can't you, Manners?" Of course I could, and he took me down to the office. So successful was I, that I became the draughtsman for the Company, at an increased salary.

During these days the boom was at its height, it was a wild financial orgy. Mine syndicates were being floated every day. There seemed to be unlimited supply of money. It would be whispered abroad, as a great secret, that Captain Piper and party had something coming on, and those who were up early would see about half a dozen drive off in a drag with horses. It was all a great secret; all were on tip-toe with anxiety, trying to find out all about it.

Next morning would be seen in a Broker's window, a beautiful heap of specimens of carbonates of lead, chloride of silver, and other minerals, which had been most probably obtained from the Broken Hill Proprietary Company's lease, and a ticket stating that the subscription list for the War Dance Silver Mining Syndicate would be opened at 10 o'clock tomorrow morning, 1000 shares at £5 each, £2/10/- on application, and £2/10/- on allotment.

There would be no report, not even a statement that the specimens came from the War Dance Lease, nor even that the lease had been granted. But long before 10 o'clock in the morning there were hundreds of men awaiting the opening of the office. By 12 o'clock the list would be closed, and an announcement put in the window that the shares were four times over applied for, that in a day or so it would be known how many shares each applicant would obtain.

The application receipts could be sold for from £5 to £10, and when the scrip was issued the shares in some cases rose to £20 or £30. Elliot's verandah was the stock exchange, and touts walked about calling their business. "I'm a seller of War Dance." "I'm a buyer of Last Chances." "I'm a seller or buyer." was roared all over the place. If you wanted to buy or sell, you went to one of these fellows, arranged the price, then paid over your money and got your scrip, or perhaps only a Broker's receipt for the application money. Some smart folk made a lot of money, but the mugs, the inexperienced optimists, always wanted more; when their shares were £5 they wanted £10; when they rose to £10 they wanted £15 and so on. They held on until they got their first and final dividend of 2/6 per share.

I knew a man named Thomas. He was a Surveyor and mostly taken out with these parties of alleged prospectors to mark out and locate the ground. In payment for his services he would be given a certain number of shares in the flotation. Thus he got quite a box full of scrip.

He came to me one day and said he was going to Ballarat for a holiday, because he had made his fortune.

"How much have you made?" I asked.

"Oh! Lord! I haven't had time to count it - about £40,000 or £50,000."

"Is it in cash?" I asked.

"No, it's all in scrip."

"Well, why not sell out and get the cash?"

"No fear! It will be worth double that yet."

"Oh, rot!" said I, "Surely £40,000 is enough! What more do you want? You will be right for life."

"No, no, I'm going the whole hog, I'm after £100,000'.

He went for his holiday, staying three months. He was fêted in Ballarat as a Broken Hill Mining Magnate who had made his fortune. But when Thomas came back the boom had burst, his box full of scrip was quite unsaleable. If he got £50 in first and final dividends he did very well indeed. I asked him if he had read the fable of the boy and the nuts. He said he hadn't.

We soon had to enlarge our premises because my old friend, Jack Dobson, a Phoenix Foundry shopmate, Cyril Hylton, and my brother Charlie arrived. We had also to put a floor in for the dust was infested with fleas that crawled up our legs and swarmed all over us. In these hot climates, they seem to live only in the dust, so the floor freed us entirely from this pest. The dust, too, was a thing worth writing home about.

The Silverton Tramway Co. had at this time been finished to the town, and the station was near our camp. The line was subsequently run right on to the mines, but at this time all the goods and machinery were conveyed from the Railway terminus to the Mines in wagons with horse teams. Soon the roads were cut up so that the dust was from 6 inches to a foot deep and, there being no footpaths laid out, the roads extended right to the doors of the buildings.

It was amusing to see newcomers leaving the station and tip-toeing across the road trying to avoid the dust and keep their nice shiny boots and fancy socks clean. Because this was absolutely hopeless, they soon just ploughed through it and washed their feet when they got home. Water was not too plentiful, and had to be carted several miles from Steven's Creek where there was a permanent sand soak that held water from evaporation all the year round.

There were at this time no women at all in the place, and thus I obtained my first experience at nursing. Cyril Hylton took Typhoid Fever. I think it must have been in a mild form, or Cyril would not have survived the very inadequate nursing he received. He was a contributor to the medical fund on the Mines. All employees paid 1/- per week, married men 1/6d. The doctor received this and, in return, was supposed to give medical attention and all medicines. His income from this source must have been between £2,000 and £3,000 a year, and should have been adequate to ensure the very best attention. So Cyril had medical attention. That is to say the Doctor came once a day, felt his pulse, took his temperature, and left a supply of quinine, with instructions as to how much to take.

Nurses, any women in fact, were quite unprocurable, so the nursing fell on me. I did not have to go to work until 9 o'clock, so I rose early, fed Cyril on milk, sponged him all over, and attended to all that was necessary to be done before I left. He was then left unattended until 12 o'clock, having a glass of milk on a chair beside his bed. Then I gave him another half dose, and in the evening after 5 p.m. he was again under our care. I was often up several times during the night attending to him.

This lasted several weeks, and but for Cyril's discomfort, I rather enjoyed the experience I got another taste of that medical desire for money without service.

The Doctor was complaining of the expense of running his business and in discussion with him one day, he said, "Do you know? I have used a whole ounce of quinine this month." Then again he remarked, "Take your friend Hylton. I have attended him for six weeks for a paltry 1/- per week. Why, if he had been a private patient, I would have had a bill against him for £20."

"Yes, dear Doctor, and you would have charged him £19/19/- too much for all the good you did; 1/- would have covered the cost of quinine and, in any case, Hylton recovered by the strength of his own constitution and the wonderful recuperative power of nature I believe, quite unaided by your pulse-feeling, temperature-taking, and quinine. You do not complain about the hundreds of men who pay their 1/- per week from year's end to year's end and never see you, never require you and, perhaps, if they did, have not sufficient confidence in you to call you in."

Shortly after this a woman arrived in Broken Hill. It was very amusing to see the men stand and gaze at her as she passed along the street, just as though she were some strange animal they had not seen before. It was a very welcome sight, the first sign of civilized life coming to the district.

The men were a rough lot, tough and hardy pioneers who are the first to rush to places of this sort, facing hardships of all descriptions, scarcity of water, fevers and barcoo and other ailments. But they were honest and generous to a fault. They would share their last shilling or their last loaf of bread with you.

There was no need for locks on doors, and there were very few fences. One just left the axe and other tools outside, and found them there on returning home. One thought nothing of going to work in his working clothes, and leaving his week's wages in his best trousers pocket, hanging on the wall in an unlocked camp.

There was scarcely a day went past without a fight between two men in the street. The cause of these fights was mostly some silly quarrel started through drink. In these cases the police made a ring and saw fair play. If they arrested the culprits they would soon be at the end of their resources, because the gaol accommodation consisted of a huge log to which the prisoners were chained, there being no cell at that time. This log was very seldom tenanted, and then only by some poor devil in the D.T.'s, chained up for his own safety.

I remember we organised a race meeting in these days and, of course, all the crowd was there. I counted no less than nine fights that took place that day. In each case the only two policemen acted as ring makers and referees. Oh, they were good sports and, under the circumstances, it was the best and wisest way to settle their differences.

It was in these early days that S.H. Holly came. He was a real Yankee from Salt Lake City, a concentrating expert. He was a practical man only, but a man of the most wonderful resources that I have ever met. I was delegated to him as draughtsman, and we became very intimate. He had a wonderfully strong voice, and prided himself on being able to make his voice heard over the roar of the machinery. His vocabulary, too, was rich and rare. He could swear, and use more profane words to the sentence than anyone I ever heard. It was not with vicious intent. It was a part of his normal conversation, and the presence of ladies did not make any difference. I think he was quite unaware of his profanity.

There was another Yankee, named Harper, imported as Underground Manager. He was pretty proficient, and eloquent in profanity, but had to take second place to Holly.

One day Mrs Patten, the General Manager's wife, reprimanded Holly for using such bad language, and Holly replied, "Me swear? By the lovely Jesus Christ, I don't swear. You ought to hear that B..... B...... Harper swear." Mrs Patten gave him up as a bad job.

Under Holly's instructions, we designed five large concentrating plants with Collom Jigs and Cornish Puddles or slime tables, and large steam stamps. There was one on the Broken Hill Proprietary, one on the B.H. Block 14, another on the British B.H.P., and another on the B.H. Junction, afterwards the Junction North. The total cost of these various Mills would probably be £250,000, and they were practically a failure. Holly confided to me later on that, when he arrived and discovered the problem he was set to solve, he told the management it was beyond him.

He had been accustomed to the concentration of copper ores, principally on the Calumet and Hecla Mines where the problem was the separation of the copper ores with a specific gravity of from 5 to 6, from gangue with a specific gravity of two and a half. This was a simple mechanical problem. But the

WGM's contemporaries at Broken Hill c.1887.
E.G. Bristow (LHS) H.H. Schlapp (seated)

A.E. Savage, E. Bull, E.G. Bristow.

Broken Hill problem was the separation of zinc ores from lead ores with a difference in specific gravities of about .2. It was almost impossible. The management urged that, because he had been brought out to do the job, Holly had better make the best of it. Thus is experience gained, and money wasted. But I suppose this is the first time the shareholders have been told that the companies spent over a quarter of a million to educate their managers. And my subsequent experience in Mining engineering accentuates this fact as the relation of facts will demonstrate. I have seen millions of pounds wasted on mining plants through the ignorance of the management, or sometimes through what is known as swelled head.

Holly could have solved most problems, but this was too much for him, and was too much for many of his successors, until the flotation process came to the rescue. Holly never claimed to be scientific, he was purely mechanical, and designed some very clever machines, one being the saw for cutting out the square set system timbers that did the work of 20 men on the B.H.P.

With all his roughness, Holly was good and tender hearted. I think he was homesick at times, and often talked of his wife and family "way back in 'Murica'." When writing home one day he looked up and asked, "How do you spell 'quickly', Manners?"

I told him. Then he said, as a kind of apology for asking, "You know, Manners, I had no education. I left school when I was eight. All the education I ever got was damned hard knocks."

He was wonderfully quaint in his expressions. One evening after a very hot day, and near the end of a three month's spell without rain, it was looking much like a thunderstorm. There were no sanitary laws then in Broken Hill, folk just threw their rubbish out in the back yards. Holly felt like a yarn so, at 5 o'clock when I was knocking off, he said, "Are you in a hurry, Manners?"

"No," I replied.

"Well, let's have a smoke and a yarn."

So we got cigars and sat on the front verandah of the office, overlooking the town.

"Guess it looks like a b....y thunderstorm."

"Yes." I replied, "I think we'll have rain at last."

"Well, look here, Manners, if the rain comes down on all that b....y garbage lying in those b....y back yards, and the b....y stuff begins to fester, up goes the bloody angels."

Holly was quite serious, but this quaint method of expressing the result of unsanitary conditions was too much for my powers of control, and I laughed. I think Holly wondered why I laughed. It was so serious a subject.

It was while I was a draughtsman for Holly that I met Captain Morish. Mr Sam Wilson brought him into the office and introduced me to him as the son of Mr W. Manners, Mine Manager of Ballarat. Mr Morish was the Manager of the Broken Hill Central Mine, and knew my father well. He was very pleased to see me, and disposed to be friendly. I had not yet got over my bashfulness, when in the presence of the leading citizens and, in consequence, had nothing to say for myself, and found it very hard to make conversation.

I did not follow up the acquaintance as any young gentleman should have done. I know now I should have called on Mr Morish, perhaps some Sunday afternoon, and joined the circle of his friends, but I had not the good sense or breeding to do this. I never saw Captain Morish again but, imagine my chagrin when I realized what I missed. They had not at this time discovered Silver on the Central Mine. In fact a Cornishman from Moonta, for Moonta was the Australian Cornwall, informed me that he "would eat all the silver that coomed out of the Central Mine."

The shares were then 2/6d. each. I remember I bought 100 at 2/6 and sold them at 5/-. Of course, this was just pure chance. They had discovered the Broken Hill lode, and the shares went straight up to £10 within three weeks. Captain Morish resigned, to retire with a fortune of between £40,000 and £50,000, I learned that all his intimate friends had been in the know, and had made big rises. Oh, ye Scottish parents, what an inheritance you give your children when you relegate them to the background in your society. And what a price the poor young people pay to gain that standing in society that you should have handed down to them.

This was only one of many things I missed through that lack of training which should have let me know that I was as good as the best of them, if not much better than most.

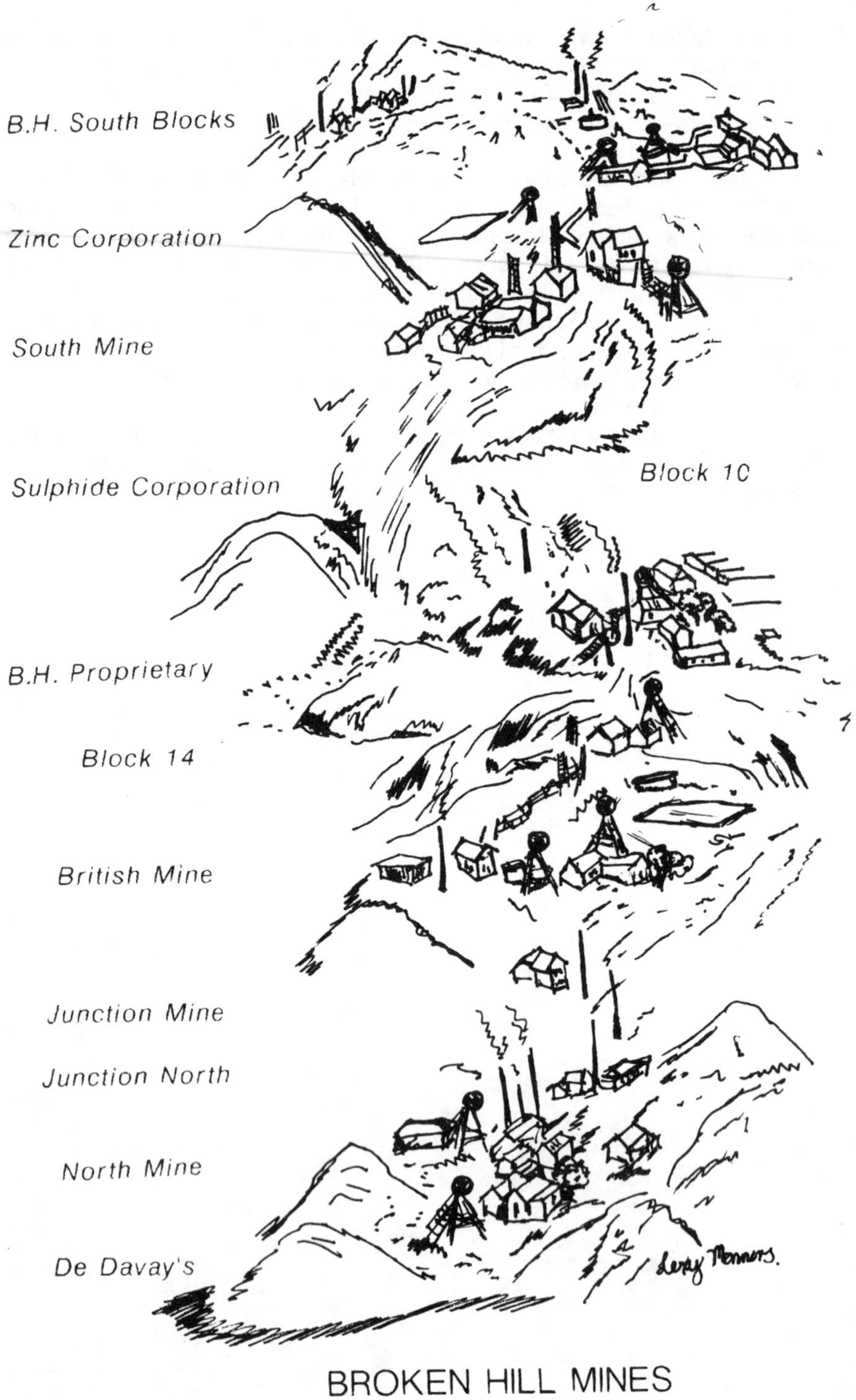

BROKEN HILL MINES

8

Fires And Strikes (1888)

Water being one of the luxuries of Broken Hill, the Fire Brigade had to resort to other means of trying to extinguish any fires that occurred.

The buildings were of such a flimsy nature at this time that, in the event of fire, very little could be done by the firemen.

Sometimes shovelling the ever present dust onto the fire saved some small part of the premises and, in the event of other buildings being in close proximity, wet blankets would be hung on the walls and roof to prevent the spread of the flames by conduction of heat and flying sparks.

The big fire, however, taxed the genius and energy of the brigade, and they displayed some amazing brain deficiencies. It was on a hot day on the fifth of November, and was apparently a joke being perpetrated by Fate to commemorate the anniversary of Guy Fawkes Day.

The whole street of buildings about a quarter of a mile long were of the temporary type until we reached Elliott's Denver City Hotel at the far end.

The fire started near the end of the street farthest from Elliott's and spread with amazing rapidity. Soon salvage workers and firemen were hard at it rescuing goods from stores, and piling them up on the middle of the road.

The Fire Brigade Captain conceived the idea of pulling down a building further along the street to create a gap which might stay the spread of the fire. So soon, with ropes and axes, a large body of workers was busy about three doors from the fire, chopping and pulling away at the building. But before they had it half down the fire would be on them, so they would rush a few more doors down, and devote their arduous labour to demolishing another building, but with the same result. The fire would overtake them long before they could make a gap.

We got a shock at one shop as we rushed to start on a new gap making venture. We started to run into a shop to be met by a man with a revolver in each hand, and commanding, "Stop! If anyone enters here, I shoot."

We did not enter, but started on the next shop, alas, with a similar fate; too late, the fire was on us again. The man with the pop-guns was quite right, he was well insured and, if his stock was burned, that was all right. But if it were hauled out and scattered over the dirty streets, trodden and destroyed, the Fire Insurance Co. would have the smile.

So the fire went on its merry way until it reached Elliott's. This was where the firemen's great presence of mind and brain power was revealed. Elliott's was a two storey place with a balcony over the footpath, with marble top tables and iron stands arranged around for visitors. The place was quite safe. Wet blankets were spread on the roof, and over the windows, and the brick wall would stop the fire. Jack the genial landlord stuck to his counter selling his liquor quite unconcernedly, and as fast as he and his assistants could serve the drinks.

But the firemen rushed in. One darted to the corner of the balcony and chopped down the flag pole, while others, with great intelligence, threw the marble top tables over into the street where, of course they smashed to atoms. Others rushed into the bedrooms threw down mirrors and crockery which met a similar fate.

It was not until someone more calm than the professional fire-fighters went upstairs and stopped them, that these fellows were induced to calm down and adjourn to the bar and have a long beer.

This was a great disaster at the time, but was eventually for good because decent buildings replaced the shanties that had served their turn.

The next great excitement was the strike. The Union leaders had decided that all employees on the Mines must belong to a Union.

This was forty years ago, but unions are still fighting and striking for the same purpose. It is a strange trait in the strike leader's mind that everything is an interference with the liberty of the subject, but not the compelling of everyone to become a unionist.

If a sincere person wishes to reform or restrict the liquor traffic, he is interfering with the liberty of the subject.

If one proposes to abolish betting and other forms of public gambling, he is again encroaching on that much desired liberty. And the ones who cry loudest about these encroachments on liberty are the first and strongest advocates of the imposition of the compulsion on all workers to belong to a union, and contribute to the funds for enhancing their political aspirations, and supporting a lot of lazy union and strike leaders who are too cunning to work, but know how to make those who do work support them. So this strike was organised, and the Mine picketed by hundreds of men, who absolutely refused to allow anyone to pass on to the Mine.

It was the 4th of July. Mr Harper had gone down town to get his dinner. He had been allowed to pass off the Mine but, on returning after his dinner, was stopped by the pickets who informed him he could not go on to the Mine.

"But", said Harper, "I'm the Underground Manager."

"We don't care who you are", was the reply, "our instructions are to allow no one on the Mine."

"What am I to do then?" asked Harper.

"Go back to town," he was told.

"Well by ....," said he, "I guess I'm in Australia, and I suppose it's the way you do things here, but .... I don't like it, by ...... I don't, especially on the 4th of July."

This tyranny lasted only a day or two, after which the officers of the Mines were allowed to go and come as they liked, and were all sworn in as special constables to guard the property.

The Miners took the matter philosophically and, although there were lots of quarrels, and much speech making, the strike dragged along for about three months. In spite of assistance from other centres, strike pay became scarce and, as most of the families began to starve, the strike fizzled out, and the men went back to their work, while the union bosses began again to organise and obtain funds to continue the fight at some future time.

I was contemplating marriage about this time, in a little nest in the shape of a wood and iron four-roomed cottage. Girls were indeed plucky who were coming to so rough a place as this was at the time, with its poor buildings, and everlasting dust. But there were plenty of pioneer girls, as well as boys and men, and I have met them away out in the back-blocks where there was only the one white woman within fifty miles. They were happy and content having a devoted husband and dear children to cheer their lonely life.

I was looking around for furniture, and the chance came of getting a rather good supply. There was a man called Edwards, who was an accountant on Block 14 Co. He had been a Bank Manager in Adelaide, but had too frequently, "Put an enemy in his mouth to steal away his brains." This fault, though perhaps more prevalent in Broken Hill, in these early days was not regarded so seriously as it would be in a big city; so long as one did not neglect one's work, he could get drunk as often as he liked.

As in all new places, a large number of liquor licenses had been granted. There was a drinking saloon at every street corner.

Edwards had left the office at 10 o'clock with a cheque for £300 to get money for office petty cash from the Bank. At 12 o'clock he had not returned. Edwards' boss was getting quite anxious, and another man on the staff and I were sent down to hunt for him and, if possible, bring the cash back. We first called at the Bank and ascertained that the cheque had been cashed. Then, knowing his weakness, we searched the pubs. There was no thought of Edwards having absconded with the money because no one could get away from Broken Hill except by train, unless he were equipped with a camel team.

We soon found him. He was standing at a bar with about half a dozen loafers, and he was very drunk. The leather bag containing the money was reposing on a form in the bar some distance away from its caretaker. We secured this, and tried to induce Edwards to come along with us, but he could not be shifted. So I sent my friend back to the Mine with the bag who, to our great relief, found the money correct. He had evidently been discreet enough even when drunk to keep his mouth shut about the money, and the very careless and casual way it was neglected in the saloon would divert suspicion as to its value.

Edwards would not face home or the Mine, and I had a difficult time with him. The biggest struggle was to get him past saloon doors. This is the greatest menace to all these poor devils who so easily lose control of their senses. The door stands open invitingly, and the greedy saloon keeper will sell drink to a man so long as he has the money to pay for it. I believe the majority would take a poor drunk's last sixpence if they knew his wife and children were starving at home.

However, I got him to the police station. The police were a good, sensible lot of men and, with their assistance, we got Edwards' consent to lock him up if we would provide him with a couple of bottles of beer.

So he was locked up for the night. I called for him in the morning, and he returned to his much distressed and unhappy wife, a sad and miserable fellow. Of course he lost his job, and had to leave Broken Hill. I believe he finished up in the humble occupation of driving a cab in Adelaide.

Such is the fate of thousands, hundreds of thousands, of alcohol's poor victims, who are forced down and down the social scale until they reach the very dregs of society, and thus help to build up the slums that disgrace our modern civilization, and destroy the beauty of our large cities. A smiling Government living on the luxuries of the land provided by these monopolies and vested interests, keeps on smiling, and permit this stuff to be sold at every street corner. Meanwhile, fawning church Bishops, Archdeacons, Vicars, and other churchmen look on, tolerating its nefarious influence, while they drink their champagne bought with the dividends of their brewery shares, smile, and offer up public prayers for temperance.

Argent Street, Broken Hill c.1887.

Remains of Argent Street buildings destroyed by fire 5 November 1888.

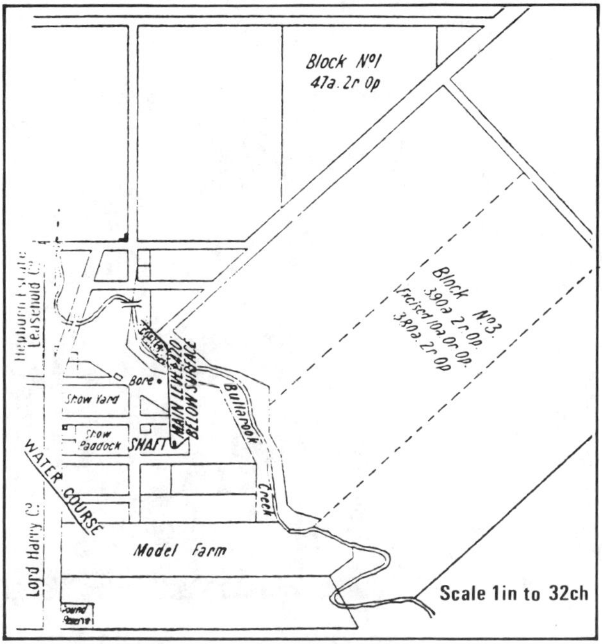

Plan of Smeaton Reserve Mine

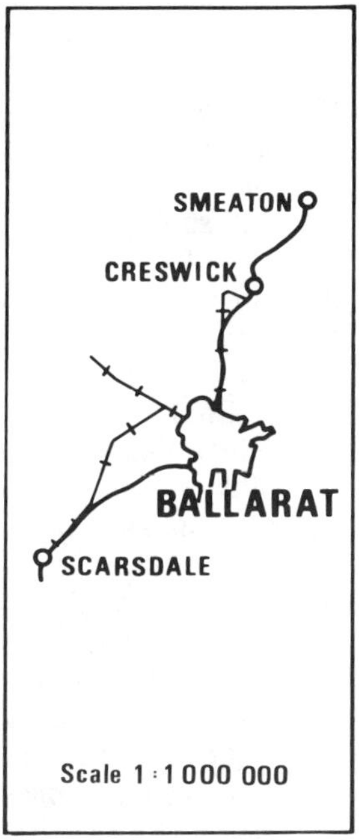

Smeaton

During W.G.M.'s time in Broken Hill, his father William moved from Ballarat to Smeaton, where he managed and reopened the Smeaton Reserve United Mine.

Previously, in 1882, the company had been formed and secured a lease on private property in the township of Smeaton, on the eastern side of the main Ballarat to Castlemaine Road. The shaft was sunk on the western bank of Birch's (Bullarook) Creek, at the rear of the Smeaton Showground.

Operations proceeded smoothly until April 1884 when the workings struck a water flow, well beyond the mine's pumping capacity.

Work ceased until William Manners arrival in June 1886 when 2 x 16 inch draw-lifts were installed, pumping 7 x 7 foot (ft) strokes per minute. The shaft was de-watered and shaft sinking resumed on September 15 that year. An additional pump was required a few days later, to bring the total pumping capacity to 60,000 gallons per hour.

After sinking through hard rock from 220ft depth, after some black and grey clay, a seven foot intersection was made of gravelly wash with heavy waterworn quartz boulders containing uneconomic gold.

No payable gold was discovered until sinking to 420ft and horizontal driving of 1,535ft was completed.

By October 1889 an area of wash 600ft x 400ft was opened up.

A total of 189 machines were washed for a yield of 712oz 15dwt of gold averaging 3oz 15dwt per machine.

Driving continued but with little success and after months of disappointment the operation ceased in March 1890.

Charles Brown was an early Ballarat pioneer. Born in Airdrie, Scotland, but unlike his more famous brother John Brown, the Clyde ship-builder, travelled to Australia and Ballarat where he commenced his foundry in the early 1850's.

He was smelting iron in Main Road (near the Charles Napier Theatre) by the mid 50's, before opening the Albert Foundry in 1862, with his brother Andrew (later a director of the Band of Hope Mine).

Charles Brown then opened up at the rear of the Albert Foundry, in Grenville Street, later shifting to Grant Street, just west of the Yarrowee Creek, near the Sir John Franklin Hotel.

The Grenville Foundry grew to be a large concern, designing and manufacturing mining and other machinery of every description, from boilers to steam engines.

Charles Brown 6.5.1826 - 16.2.1887

An extensive fire burnt out the pattern shop and did other damage to this foundry about 1880; but they carried on until 1887, when Charles Brown died at the age of 60. He had thirteen children and his eighth, Margaret, married W.G. Manners at Ballarat in 1889.

9

Marriage And More Doctors (1889)

Having my nest in "pretty good shape" as my friend Holly would have expressed it, I journeyed to Ballarat to "enter into that bourne from which no bachelor returns." Of course, at this critical period of a person's history, one is liable to feel a little excited. It is a great undertaking and, I think, a great deal more so for a woman than for a man. Of course for her it is the climax of her aim and hope in life. She is made to be protected, petted, and loved, and, although a great many of them miss the mark by marrying the wrong man, the great majority are blessed with happiness, peace and contentment, even through poverty and strife. It is a sure thing that if love is the foundation of the marriage, and commonsense remains in the home as love's companion, all will be well.

My brother-in-law to be was a draper, littleman and a clothier, so I was immediately taken in hand and polished up. I suppose I needed it very badly for with the dust and dirt in Broken Hill one was not disposed to be extravagant in dress, and doubtless patronised the ready made counter, not having the refining influence of a woman's good taste in clothes. The "reach-me-downs" were probably not in the same class as would result from a Bond Street tailor.

However, I was fitted out with a new suit, Chesterfield overcoat, boxer hat, kid gloves, cane, etc. I did not recognise myself in the large mirror, and was in doubt whether my mother would remember me, and whether she would give a thought to my moleskin pants period.

I had decided to spend a week end at my parents' home which was at this time on the Smeaton Reserve, at Smeaton. The train left Ballarat at about 9 o'clock on Saturday night. My fiancée and two or three of her sisters had come along to see me off, and my brother had come along to go home with me. He was being seen off by two male friends. One of these was little Harry Larner who drove the Rev. J.W. Ingles' coach. He was a funny looking little chap with a snub nose and a half open mouth that showed irregular and buck teeth. He was about forty years of age, about five feet four inches high, and talked with an impediment. So he had numerous graces to recommend him to the attention of the interested fair sex.

We had a pleasant week end in spite of the very cold and wet weather. We had to do several miles per coach after leaving a station called Allendale. On my return I was to bring a crayon drawing back with me. This was one my fiancee had done for me. It was wrapped in brown paper, and carried in leather straps.

On reaching the station on Monday morning, which was still a cold wet day, I found I had about half an hour to wait for the Ballarat train and, because there were no waiting room, I and the other passengers strolled up and down the platform. I suppose I looked like a commercial traveller and, being young and fresh looking, and well dressed, I was probably worth a passing glance. Walking to and fro and meeting me every few minutes was a little girl with red hair and a good supply of freckles. She was not bad looking, and our eyes met every time we passed. I knew that I had only to impart the valuable information that it was cold, for the mutual exchange of confidences to have become warm. But that would not do; I was to be married next Wednesday so, of course, I had to be good.

There was another, a very nice looking girl walking with her arm resting on that of an old man, most probably her father. We also noticed each other, but neither spoke.

At last the train came in and, to avoid any chance of meeting the interesting ladies again, I rushed for an empty first class carriage. Imagine my surprise when I looked around and there was Freckles sitting opposite me, and the girl and her father on the same seat with me.

It was soon discovered that there was a small leak in the roof. Water was dropping on Miss Freckles and, instead of moving along, she having the whole side to herself, she came over and sat beside me. No remarks were passed until we reached Creswick, when the old gentleman enlightened me as to the condition of the weather.

I remarked that I felt it particularly, having come down from a warmer climate.

"Where have you been?"

"Broken Hill." was my reply. Then while talking of Broken Hill for a few moments the train started. This made conversation with the old gentleman difficult, owing to the noise of the rolling stock.

But then Miss Freckles looked up from her book and asked, "Do you know many people in Broken Hill?"

"Oh! yes quite a lot."

"Do you know Mr. Park?"

"Yes. George Park, quite well."

Then her information encyclopaedia was opened.

"He's going to be married." she said.

Of course I knew all about this matter, but to encourage her I asked, "Whom to?"

"Miss McLarty." Then by questions and answers, she told me lots of things about this interesting pair, their family history, and a lot of statistics that were not quite complimentary.

She ventured another question. "Do you know Mr. Manners?"

"Oh! yes I think I have met him, there are two of them in Broken Hill, aren't there?"

"Yes." she replied, "Willie and Charlie."

"Yes, of course I thought there were two." said I.

Then Freckles remarked, "Willie's going to be married too."

"You don't say so. When?"

"Next Wednesday."

"And," asked I, "whom is he going to marry?"

"Miss Brown."

"Miss Brown —— Brown," soliloquised I, "Brown —— I think I have heard that name before. Which Brown is that?"

"Her father owned the Grenville Foundry, you know."

"Oh! yes, I know now. I have seen the place. And what kind of fellow is this Bill Manners?"

"Oh!" she volunteered, "He's a funny little chap with a snub nose, and buck teeth. He is very short and very plain looking.

From this description I realized that she had seen Harry Larner with my brother on Saturday night, and had jumped to the conclusion that he was the prospective bridegroom.

So I replied, "Yes, that's a very good description. How old would you think he is?"

"Oh! about forty or more."

"Time he was married." I suggested, "And how old might Miss Brown be? Tell me something about her."

This was her chance for romantic news, so she told me: "She is about forty too. She is quite an old maid, and on the shelf. She couldn't get anyone else but Mr Manners, she is so plain herself. They have been going together for years and years, and her mother told Mr Manners that if he did not marry her this year the engagement would be broken off." And so on, giving me lots of information, none of which was true, but all of which was uncomplimentary and unkind. The mother, old Mrs. Brown (a dearer, more good hearted old lady never breathed), was a hard faced, bad tempered, harridan, Mr. Manners was so afraid of her. When "Freckles" had quite finished, and destroyed the poor bride and bridegroom's character, and appearance, and we were nearing our destination, I suggested, that, because I would soon be returning to Broken Hill and would see these friends of hers, I would like to be able to tell them that I had met her, and give them any kind messages she would like to send. But, in order to do so, I would require to know her name.

"All right," she replied, "I will tell you my name on condition that you tell me yours."

The Ballarat station was very near now. Of course I agreed to this.

"Well," she simpered, "My names is Annie O'Connor. Now you will tell me yours."

We were just drawing into the station.

"Of course," I said, " I would not be mean enough to ask your name" The train stopped and a porter opened our door. We stepped out on to the platform, "and then not tell you mine. Now, Miss O'Connor, I have enjoyed your conversation, and am much enlightened, and if you will come to Grenville Street, on Wednesday afternoon at 3 p.m., you will see me again. There's my card."

As she read, 'Mr W.G. Manners, Broken Hill, N.S.W.,' she turned pale. Her dark brown freckles turned yellow. A small group standing close by saw her change colour and, thinking I had jilted the girl, remarked, "Oh! Oh! Oh!"

I, feeling that my presence would not help her to smile, turned and departed, leaving her to ponder over the foolishness of talking about people she knew nothing of to people she did not know.

The wedding went off nicely but I did not see Annie O'Connor there. She evidently did not think it expedient to accept my very cordial invitation.

I had my first experience of wife control at the marriage breakfast when I found my indulgence in the wine and whiskey, which was liberally supplied at the feast, strictly vetoed. Consequently, I got over the ordeal in an ideal state of sobriety. I found out afterwards that there was a good reason for this: a friend of ours named Lee Farr had been married a few weeks previously, and had been rather kind to himself. Perhaps his courage required sustaining. However, they left to spend their honeymoon at Queenscliff. When the train stopped at Geelong for a few minutes Lee got out to have another, and then another, and when he came out the train had gone, taking his bride away with it to fend for herself and weep on her lonely pillow. There being no other train until next day, Lee had the first night of his honeymoon with "another".

My bride wasn't taking any chances, so no mishaps occurred. Our honeymoon was enjoyed, then the arduous journey of the return to Broken Hill undertaken. This was an experience for a young girl who had not been into the wild back blocks before. But, because the train was through to Broken Hill we were spared the experience of "sphankin ponies."

Great interest was taken among my friends as to how Mr Holly would behave when he came to see us. They were anxious to be present to see if he could refrain from swearing.

So, to oblige them, I invited Holly and two or three of them to come and see us in our new home on a Wednesday evening, and Holly promised. But, on the Tuesday at five o'clock, as I was packing up my work with a view to going home, he said, "Be at home tonight, Manners?"

"Yes," I replied.

"Guess I'll come right along."

"Goodo, Mr Holly." said I. "But you know you'll have to take pot luck.'

"Guess that's O.K. I don't come along to your place just to eat." And so we strolled off. On nearing home I saw my wife in the distance coming to meet me, they do that you know, for a few months. She saw us then immediately turned round and hurried home, calling at the butcher's en route.

Holly saw her and said, "Is that Mrs Manners?"

"Yes," I replied.

He stood for a few minutes looking up into the sky then said, "Guess it's looking like we'd have rain pretty soon." He loitered and talked, taking about a quarter of an hour to do the last two hundred yards. I could not but admire his wonderful consideration and, when we arrived home everything was at its best. Holly entered quietly and, in a low voice which I had never heard before in my two years acquaintance with him, said he was very glad to meet my wife, and congratulated her on the very pretty cosy home we had. He stayed all the evening smoking cigars and talking of his home and family. And, with the exception of saying Je— once or twice, and pulling himself up in time, used no language that would wound the sensibility of the most fastidious lady.

He came often after that, but always by himself, and always the gentleman, giving us the impression that with all his gruffness and swearing he had a soft spot for home and his family, and liked our nice quiet fireside, which made him feel more at home than any other place he had been to. So time passed happily for six months or so.

In the meantime, the designing being done, I took a job as foreman of Block 14 Concentrating Mill. This was hard work, but the hardest was when the piston rod of the big steam stamp mill broke and blew the cylinder end out. We borrowed a cylinder end from the British. I worked thirty six hours straight on, having food brought to me, and got it going again. I had twenty four hours in bed, then returned to duty that same afternoon. The repaired piston rod broke again, and out went the second cylinder end. We then borrowed the spare from the Proprietary, and I worked another thirty six hours, keeping the mill going both times on the reserve ore in the bins. But this was too much for me. I soon found that I had typhoid fever. This was bad: the doctor who attended me was speculating in mining, and I was neglected a good deal.

My poor wife had had little experience of nursing at that time, and I had a really bad attack. I was allowed to have meat too soon, and had relapses stretching out the illness over five months. My poor wife was in a terrible state, because my life was frequently despaired of, and our first child was soon to be born. She was 800 miles away from her family and, as the mill had shut down a month or so after I took ill, my salary stopped. We were soon in debt to all our trades people and the doctor.

When one morning my wife was outside working and wondering what she would do, Mrs Zeb Lane[1] came along and greeting her saying, "Now don't get excited. I have something nice to tell you: someone has come to help you."

Intuition, instinct, or great love, I don't know which prompted her and, looking towards the hooded buggy in which no one could be seen, my wife said, "My mother is in that buggy," and ran towards it. And there she was, the dear old soul had sent a wire which had not been delivered, so she called at the post office, and brought it herself. She had found Mrs Lane, whom she knew, and so they came along.

If an angel had come from heaven, she could not have been more welcome, nor have done more good to us. She nursed us both with her gentle loving skill. To this day, although she has gone to her reward, I thank her for her timely arrival and loving care. 'Tis certain she saved my life and, as in other cases, I recovered by the aid of my dear old mother-in-law and my splendid constitution, in spite of medical neglect and incompetence, which characterised the whole of the treatment I received at the hands of the doctors and their juvenile understudies.

However, I recovered, but never regained my original weight and strength. I resolved, in consequence, not to return to the mines with their dangers of lead poisoning and a possibility of further illness, so started in business for myself as a consulting engineer and draughtsman.

Note 1

"LANE, Zebina. MAusIMM, MLC.
Born in Victoria at Bendigo where he began his career in mining, Lane became the Manager of a silver mine at St Arnaud. After joining the Inter-Colonial Smelting Company in Broken Hill and being the first Mayor of this city, he moved to Western Australia in 1893, and floated the Great Boulder, The Boulder Perseverance, and several other mines at Kalgoorlie in 1895. He floated the Collie Proprietary Company to mine coal at Collie in 1899, and tried without success to briquette its product in 1902. A Member of the Legislative Council from 1903 to 1908, he returned to the Eastern States in 1912. He was a foundation member of the Australasian Institute of Mining Engineers in 1893 and a member of its first council. He died in Berlin, in 1912."

...from *Mines, Miners & Metallurgists in Western Australia* By D.A. Cumming 1991.

Chairman at Prayer Meeting.-Brother Tresize will now engage in prayer. Please Brother Tresize, don't pray long, because Brother Grub's on night Shift.

10

Parsons And Sharebrokers (1889)

It was at this time that the Rev. Mr Buntine came along from Sydney for the purpose of organising a Presbyterian Church in Broken Hill. Unfortunately he managed to rope me in as a member of the new Board of Management.

I say unfortunately advisedly, because it is these kinds of men who bring the church of all denominations into ridicule, and drive the people who would otherwise have religious inclinations out of the church into the realms of scepticism.

Rev. Buntine was an ignorant, cunning, Scotsman. How he managed to become an Ordained Minister in the Presbyterian Church, which demands a high education of its ministers, is one of those problems that can only be explained by men in authority, if they would not be ashamed to explain it.

He formed the Church all right, got a strong committee together, then started fishing for a call as the permanent Minister. It was the law of the Church at that time that a Minister must not personally approach members of the Church to influence a call. But he took members into pubs, shouted them whiskey, and thus got their ears to plead his case. The members were not impressed. His sermons were such silly, childish rot that they put him clean out of the running.

Rev. Buntine would usually take one of the parables and read the whole chapter which, in itself, was sufficiently explanatory and clear, in fact beautifully concise and poetical. But he would expand it, enlarge it, exaggerate it, and make it ridiculous, so that it would be laughed at by Sunday School children. As an illustration, on one occasion he took the parable of the lost silver piece. This is beautifully expressed in two verses, (St. Luke 15. 8-9). and its application in one (St. Luke 17. 7). But this is how Rev. Buntine dealt with the subject

"Dear Brethren and Sisters, we have taken this subject of the woman who lost the silver piece as our text this evening. You will note that this poor woman had only ten silver pieces. Let us imagine for the sake of understanding her position that these silver pieces were florins or half-crowns.

Now you can understand what a loss this woman must have sustained when she lost one of her ten florins. She was probably at the door at the time purchasing some chops, or perhaps butter, or perhaps onions, or perhaps potatoes for her husband's dinner, and was wondering how much she could spare from her small store. I am sure many of you must have had the same experience. Now, as she was getting out her money from her purse, or it may even have been a stocking, one of these coins slipped through her fingers and rolled and rolled and rolled until it hid itself in some obscure corner.

You all have no doubt had experience of how when anything is dropped it rolls away, and you search, and search, and search, and cannot recover it until you find it. Oh! it is sad. So it was with this poor woman. She first of all put on her specs' then went down on her hands and knees, and searched, and searched, and searched. But, because she could not see it, she did not find it.

Then the parable tells us that she took the broom, it is not particularly mentioned whether it was the brush broom or the hair broom, but this is not really an important matter, and she swept, and swept, and swept the carpet and the linoleum, and the floor under the chairs, table, and bed, and in all the corners that had not been swept for many days. But, again, she was unsuccessful, she did not sweep it from its hiding place, and therefore did not find it. Oh! it was sad.

Then, after wiping the sweat, no I beg pardon, perspiration, from her brow with her apron, she took a candle in her hand. The parable does not say that she lit it, but common intelligence applied to this problem informs us that she must have struck a wax match and lit the candle. Such a careful woman as she was would not light it at the fire. With this lighted candle she searched, and searched, and searched. She looked under the chairs, and under the tables, under the beds, under the mats, and under the carpets, and linoleums, and at last, Oh! great joy! She looked in a rat hole in the wall and there she found the coin. Oh, what great luck! Now what did this woman do? Did she rush out with this florin and buy herself a new hat. No! Did she rush out and buy herself a new dress? No! Then what did she do? She called in all her neighbours for miles around and, holding up the two shilling piece, or, as I have previously mentioned, it may have been half-a-crown, and said, 'Rejoice with me all, you dear old women, for I have found my two shillings.' And they all rejoiced because they had not lost any money. So it was, dear Brethren and Sisters, the woman rejoiced more over the half-crown that she had lost and found than over the nine half-crowns, or as I say it may have been florins, that she had not lost, which reposed safely and securely in her stocking."

And so Sunday after Sunday these intellectual outbursts were perpetrated on a patient and submissive congregation. But when it was a question of inviting him to remain as a permanent expounder of the Gospel the Committee were adamant.

He came into my office one day and suggested I propose the call. I refused on the grounds that I was the youngest man on the Committee and had had no previous experience. He then, for my guidance, wrote out the words forming the call. When it became known that he had done this a great rumpus was caused, and he brought a lawyer into my office with a view to an action for defamation of character. Buntine absolutely denied that he had asked me to propose the call. When I produced the paper with his own writing he had to acknowledge it, after which the lawyer took him from the office, and that matter dropped.

However, he beat us. He had, with the aid of a friend, Dr Dow, worked up a clique to pack a congregational meeting, and the committee was asked to resign. They left the meeting in a body, and Mr Buntine got his call, but only stayed about six months when for the lack of a congregation and his consequent salary he was forced to retire.

My illness had been the means of landing me about £250 in debt, and business was not too brisk. So I borrowed £100 from my eldest brother and paid £60 of this to satisfy the most pressing of my creditors. I had £40 which I intended to retain to help me in business and keep my small family supplied with the necessaries of life.

Friends were very kind, and put work and opportunities my way. And I had one friend, a Mr Joseph Langdon who, like Captain Morish, had been a friend of my father's. He was the General Manager of the Broken Hill Extended Mine, a show which was not in a direct line with the other big shows, and which had not at this time found the lode. The shares were about 10/- each, and the Mine had a very good sporting chance. The Manager, knowing my straitened circumstances, honestly wishing to help me, and, I believe, sincerely thinking he was doing so, advised me to buy Broken Hill Extended's.

This was my chance to make enough money to pay off all my debts and make a fresh start in life. I remembered my regret at not cultivating Captain Morish when I might have had his advice. Now here was a man in as good a position as Morish, a good, sincere friend, a man who knew the game, a man old enough to be my father. He had a brother on the Melbourne Stock Exchange, Mr John Langdon, and he gave me a letter to him. My chance had come at last. I plunged. I bought on cover all the shares that I could with my £40, my whole borrowed fortune, and waited with a cheerful optimism for the desired result.

Of course the shares came down, and kept coming down so that at the end of three months, I was compelled to find more cover or close the deal. This was offering me only one alternative, and I closed the deal, getting in return for my £40 exactly forty pence.

This was a disaster that left me just about where I was before I borrowed the £100 and, I might say the B.H. Extended never got the lode. It was off the line. This was my first, what might be called 'big' loss in mining, but since I have found that there is a sort of fate, I am not sure whether it is inherent in man or whether it is that "destiny which shapes our ends", but there is something or someone extant with a mystic power of devining the exact amount one has in his pocket, or bank, or sock, as the case may be, and has the uncanny faculty of drawing it out. It is not like the thief who picks your pocket unbeknown to you, but it is an alleged friendly approach that draws it, actually worms it, out of you with your pleasing consent. And when the utmost farthing has departed, you feel that you have not been robbed, it has all been within the law, been done for your good, no matter if you are left to starve, and the one who has wormed your money out of you is rolling in luxury.

Thus I was again reduced to bedrock to carry on a business without a penny to bless myself with, and a pile of debts all round me. At this time the B.H. Block 14 called for tenders for the erection of a large winding engine, labor only, and no deposit required. This was a chance for me, so I went into the matter.

Mr W.H. Patten had designed three unique winding engines, double cylinders horizontal coupled with 18" diameter cylinders, and 60" stroke. This was quite exceptional. There were numerable gear wheels, and levers by means of which the steam could be cut off at any portion of the stroke from 1/8th to 7/8th. The whole affair was massive and complicated. The machine was designed on the Mine. Two of them were ordered, one for the Proprietary, and one for Block 10. The one on the Proprietary had been erected, but quite a number of alterations had to be made so that the work was very costly on account of erecting it, trying it, pulling down and altering it. However, I had been ill during this time, and put in my tender on my own estimate. There were three tenderers. When the tender opened it was found that mine was £360, the next highest £680, and the highest £960. I learned after that this latter price was what it had cost the Broken Hill Proprietary Coy to erect the engine.

Of course my tender was accepted. My opponents laughed at me and assured me that I would lose a lot of money over the transaction. I was just as sure that I would not lose a lot of money for the very simple reason that I did not have any money to lose.

Before I knew myself that I had the contract, two fitters, W. Leevers, and W. Wheelwright, came and asked me for a job. I did not know at that time that, owing to the previous strike, these men were blacklisted, and their families were practically starving. So I promised nothing. But, making sure that I was the successful tenderer, I made enquiries and engaged the two men who had erected the Proprietary engine, with the result that all the alterations were known and made before erection, paid for as extras, and the engine was completed in good time. My share of the profit, including my wages on the job, was about £120.

The Broken Hill Mines were at this time considering the question of a supply of lime flux for their smelters. There were two parties in the field for the contract. These were people who afterwards floated the Tarrawingee Flux and Tramway Company, two prospectors called Young and Leith.

The latter had named his holding the Youngaleitha Quarries, and my friend Mr Thomas was engaged by them and asked me to assist him. The distance was about forty miles. Mr Leith, who was an excellent bushman, was to be our guide.

Because the weather was hot, we could take only a limited supply of fresh meat so, after the first day, we had to live on tinned meat, or "tinned dog" as the prospectors called it, although it was all good beef and mutton. We carried only enough provisions to take us to the camp, being informed that there would be plenty of food there to carry us back.

We were to clear and mark a line going out, and chain the distances between curves back. The whole job would occupy a fortnight.

We had not got far from Broken Hill before we learned of the utter destruction caused to the country, and the ruin to the sheep squatters, by the innumerable rabbits that infested the bush. These creatures are wonderfully prolific, producing a litter every six weeks or so, with the result that, when feed and water are plentiful, they increase so rapidly as to overrun a country in a few months.

At the time of our visit it was the third month in a drought after a previous good season, and the rabbits were dying in millions. They had eaten every blade of grass, every leaf of saltbush and bluebush they could reach, and all the bark off the small shrubs as far up as they could reach. Many of them had either climbed or jumped into the small shrubs to get the bark and leaves and, in getting down, had been caught by the hind legs in forks of the shrubs, and there hung and died.

They were so plentiful and so weak from starvation that they could not move out of the way. In the shade of the mulga trees, a small shade probably not more than six or eight feet in diameter, there would be as many as 100 rabbits sitting, and one could go right amongst them before they would move. There was not a scrap of feed for the sheep left in the whole country so they were being sent in thousands to a boiling down establishment. There was nothing of a striking nature on the trip except the rabbits, and the remarkable way Mr Leith could always find water for night camping.

We shot a wild turkey on our last day outward. This was a godsend for, when we arrived at the camp, there was nothing there but flour and tea. We cooked the turkey in a camp oven, and thoroughly enjoyed it after our week of tinned dog.

But coming home tested our appetites. We had eaten nothing but dry damper and tea and, to one accustomed to meat three times a day, this is a revelation. There were plenty of rabbits, as I related, but they were merely skin and bone. Also being regarded as a kind of vermin we felt that we would rather starve than eat one of them. However, when we got to Stevens Creek, where the rabbits could get a little water by scratching in the sand, and where there was a little fodder for them, we shot a couple and grilled them. But none of us would eat them, so we continued our damper and tea fare.

There was only one more day, then home. I fancy I can smell and taste to this day, that pound of grilled chops that I ate that night. There were never chops that tasted as good as those did to a real meat hungry man.

Shortly after this it became known that the Tarrawingee people had the contract, and a company was formed called the Tarrawingee Flux and Tramway Company. A bill was passed through Parliament authorising the construction of a private railway.

I was subsequently offered a job as engineer for the Tarrawingee Flux and Tramway Company, which had a contract to supply limestone as flux to the Broken Hill Mines, and was putting in forty miles of private railway to convey the same from their quarries in which was an unlimited supply of hard grey limestone carrying as much as 95% lime. Because the company was to provide a house for me, my little home in Broken Hill could either be sold or let.

At this time a very near relation had got married rather prematurely and, not having much money to start a home, but being infinitely better off than I was, being free from debt, I, in a very weak moment, offered him the use of my house rent free for twelve months, after which I would require him to pay a reasonable rent. My friend was very thankful and occupied the house for the stipulated twelve months. When I called in, at the time wondering how low I could make the rent for my friend. I was greatly shocked when they informed me that the house was not suitable to their requirements, and they had decided to shift. They went into another house and paid 25/- per week, and never even said thank you for my foolish generosity.

How long it takes some kind-hearted, generous minded old fools to learn the selfishness of the world. I found then, and have confirmed it greatly to my loss, that kindness and charity to relatives is always unrewarded. They take all you can give them, then when you have given all you can, and are compelled by very necessity to stop, you are a mean selfish cad, who has never done anything for them at all.

OPENING OF THE TARRAWINGEE TRAMWAY.

Mr. J. S. Reid's Special Train FROM ADELAIDE TO BROKEN HILL AND RETURN.

TIME TABLE

For the information of the Passengers.

A. G. PENDLETON, General Traffic Manager.

ADELAIDE, JUNE 3, 1891.

Menu

SOUP.

Printanier

Prawns in Aspec. Lamb's Tongue in Aspec.

REMOVES.

Boned Turkey en Bechamel.

Roast Turkey. Roast Chicken.

Roast Duck. Roast Goose.

Saddle of Mutton.

...ue. York Ham.

Dressed Salad.

ENTREMETS.

Bavarian Pudding.

Jelly. Port Wine Jelly.

Fromage of Apricots.

...ues à la Cream.

Tipsy Cake.

...y Pastry.

...sert.

...e.

...s.

Chateau Lormont.

...dseick Monopole.

Special...

F. D. BEACH & SON. CATERERS.

Melbourne to Broken Hill. 122

Special Train 6th JUNE, Sleeping Berth included.

Broken Hill to Melbourne. 122

Special, Ordinary or Express Train. Available to 17th JUNE, Sleeping Berth included.

ENTREE CARD 122

Tarrawingee Banquet

SKATING RINK, BROKEN HILL,

June 9, 1891, at 8 p.m.

Mr John Corlis C E

With J. S. Reid's Compliments.

...1891. ...to Broken Hill.

...ide	dep.	6·50	a.m.	
Gawler	arr.	7·28	"	
"	dep.	7·29	"	
Hamley Bridge	arr.	8·3	"	
" "	dep.	8·6	"	
Riverton	arr.	8·35	"	Breakfast [here.
"	dep.	9·5	"	
Burra	arr.	9·53	"	
"	dep.	9·58	"	
Terowie	arr.	11·0	"	
"	dep.	11·10	"	
Petersburg	arr.	11·45	"	
"	dep.	11·50	"	
Yunta	arr.	1·50	p.m.	
"	dep.	1·55	"	
Mannahill	arr.	2·50	"	Dinner [here.
"	dep.	3·20	"	
Mingary	arr.	5·5	"	
"	dep.	5·8	"	
Cockburn	arr.	5·50	"	
"	dep.	5·55	"	
Broken Hill	arr.	7·10	"	

WEDNESDAY, JUNE 10TH, 1891.

Broken Hill to Adelaide.

Broken Hill	dep.	8·40	a.m.	
Cockburn	arr.	10·0	"	
"	dep.	10·5	"	
McDonald's Hill	arr.	11·20	"	
" "	dep.	11·23	"	
Mannahill	arr.	12·35	p.m.	Dinner [here.
"	dep.	1·25	"	
Yunta	arr.	2·25	"	
"	dep.	2·30	"	
Petersburg	arr.	4·45	"	
"	dep.	4·50	"	
Terowie	arr.	5·25	"	Tea here.
"	dep.	5·55	"	
Burra	arr.	6·55	"	
"	dep.	6·59	"	
Riverton	arr.	8·12	"	
"	dep.	8·15	"	
Hamley Bridge	arr.	8·45	"	
" "	dep.	8·45	"	
Gawler	arr.	9·17	"	
"	dep.	9·20	"	
Adelaide	arr.	10·0	"	

11

Tarrawingee (1891 - 1893)

Tarrawingee was a joyous place during the construction of the railway and the erection of the machinery. Messrs. Baxter and Saddler had the contract for the railway. They employed a large number of navvies.

We were opening up our quarries, and also employed a large number of navvies. These were mostly Irish and Italian. Our foreman navvy was a Scotsman with a we' bit authority. We put it to him one day that the Scots were clannish, and that he was not loyal. Why didn't he employ Scotsmen instead of Micks and Dagos?

Scotty's reply was characteristic. He said, "Weel noo, its this wiy. If I employ Scots, when I swear at them, and tell them to hustle, they tell me toe gang toe hell, and walk off the joob. But wi' the Pats and Dagos, I can swear at them, bully them, kick them an' they juest wark a' the harder."

To the unbounded joy and delight of the landlord of the only pub in the place, Harry Banfield, the contractors and we paid fortnightly on alternate Saturdays, so that one week our boys treated the contractor's boys to a howling drunk, and the next week the contractor's boys returned the compliment, with the result that, by Tuesday morning every week, the camp was broke and Harry was waxing fat.

A little chap called Crisp was the overseer of the quarry department. He was an unknown quantity, neither Scots nor Irish or, if there was any of that normally white blood in him, it had got badly mixed with some nigger blood. He had neither brains nor education, and possessed only one characteristic of the true born navvy, which was discovered when he ultimately got the sack. That was that he cleared out leaving a lot of petty debts behind him. No navvy ever left without cheating his boarding house keeper out of a fortnight's board.

Father Connolly always came out every week in these days to give absolution to his erring flock and, incidentally, to collect as much of the money as could be rescued from Harry. It would have been a fearful thing if these swearing, fighting, drunken blackguards should have died, say on Tuesday, with "all their imperfections on their heads," and gone to Purgatory leaving no money behind to bribe the Lord to let them get to Heaven. Whereas, by Father Connolly's goodness in coming out they could paint the town red and, if they had enough money left from their spree, or could borrow enough to pay the good Father, he could wash them whiter than snow. Then if they died, they could get a harp and make the miraculous change from hard-faced roughies to beaming angels for about 5/-

The good Father was very popular with all of us. At this time our houses were not built and we had to live in the pub. We had a private dining room. the Father would join us during his visits and, in the evening, a few business people would come in to spend the time with draughts, chess, and cards. One evening the Father said, "I think gintlemin, that we might just have a little drink. I don't drink meself, ye know, but whin ye are stopping in a place loike this, one has to spind a little money, ye know. So phat will ye have?"

The landlord was called in to take our orders. Some ordered beer, others wine, others whiskey. "And what will you have Father?" asked Harry.

"Well now," said the Father, "I'll just have a little lemonade and pippermint out of the same bottle, ye know."

How he got lemonade and peppermint out of the same bottle, or whether there was a "same bottle" that held the priest's favorite, was a secret between the landlord and the Father.

When the railway was completed and our machinery erected, payday came along only once a fortnight, and the Father's visits were reduced accordingly. It is marvellous to notice how the ministrations of the Lord synchronise with payday.

Baxter and Saddler gave a spree at the opening of the railway. I will only relate one incident in connection with the affair because it bears relation to one of our human idiosyncrasies:

After the banquet which was held in Broken Hill, the participants were invited to run out to Tarrawingee in a special train to christen the line.

One of the visitors was Mr H.H. Schlapp who was at that time assistant manager of the Broken Hill Prop. Co. He and I were chatting when a steward came around with a tray containing a bottle of champagne and several glasses, and offered us a drink. Of course I took a glass. But, what do you think? Mr Schlapp refused. I was surprised and, turning to him, asked, "Are you a teetotaler, Mr Schlapp?"

Mr Schlapp was a Yankee, he talked through his nose in that peculiar Yankee way, and rather drawled his words.

"No," he replied, "I am not a teetotaler, but I'm damned if I can see the sense of drinking champagne today, and having a headache tomorrow."

Oh, ye tipplers! Just swallow and inwardly digest this reply, then ask yourselves how many headaches you have purchased with hard earned money. How many healthy luxuries you have done without in order to buy a headache. I am not blaming you particularly, I only look back with regrets, trying to calculate how many headaches I have bought and paid for both in money and misery.

With the exit of the railway men we settled down to become a quiet and respectable community of about 400 people. And, much to the regret of Mr Banfield, the drunken orgies happened only once a fortnight, although he was still accumulating a fortune at the expense of the hard-working navvies and their Monday morning headaches.

Then we began to arrange sports. The Company built us a tennis court, which was a great source of amusement for the staff. But for the public we had other sports. We obtained a cricket set and had matches on Saturday afternoons. Our manager, Mr Stirling was a fairly good cricketer, both at batting and bowling. But it was

H.H. Schlapp, Broken Hill c.1887.

not one of my accomplishments. However, because it was a difficult matter to raise two teams, I was always picked to play. I was generally about last man to go in. My captain would get my ear as I was going out to bat and solemnly request me not to try to score. "Just keep your wicket up, block everything, and let the other fellow score." This was no doubt cricket wisdom, and would probably work very well with a practised player. But not with one who was not well acquainted with the vagaries of a cricket ball, and could not understand why a ball pitched two feet to the off or on, could upset the middle stump. His advice was absolutely lost. However, it was my duty to obey instructions, so I tried this game. But, alas by some process of bowling legerdemain, my bat would not be where the ball ought to have been, and my stumps were soon in the same predicament. I was out for a duck.

One day we got up a match, married against single. The single men had been in and put up a score of ninety six. I was ninth man in; our score stood at sixty two with only one wicket to fall. The bachelors were jubilant. The married men wondered how they would be able to face their wives and families with such a defeat to their discredit.

I got the usual injunction from the Captain, "Keep your wicket up, old man. Let Brown score", etc. etc.

As I walked out to the field I soliloquized, "Hang this silly nonsense of trying to keep an unstable wicket up. I'm going to have a game of cricket." So I squared my shoulders and hit out with all my might. I knocked the ball all over the field, skyed it again and again, but always, with luck, out of reach of the fielders. They tried fast bowls, slow bowls, googlies, and all the strange bowlers' arts. But it was all the same to me; I hit them. I knocked up thirty two runs while my partner scored three, then was caught out. I had won the match and saved the faces of the benedicts. I was cheered as I returned to the pavilion, and congratulated on my great innings.

"You never gave them a chance." they said. They did not know that this was pure luck. I just hit promiscuously. I did not study the field, but had won my reputation at last, and that by disobeying orders.

The next great excitement in Tarrawingee was a race meeting. No town in Australia, no matter how small, is worthy of its name unless it can run a race meeting. Blood horses are not necessary, any old crock, bakers' and butchers' delivery horses, will do in an emergency, so long as the people can congregate, make bets, get drunk, and be in fashion.

There was one squatter in the district. a young English Oxford graduate named Lewis, who kept some good horses. It was a difficult matter to handicap them so that he would not secure all the prizes.

Everything went well until the steeple-chase. This was a race of about two, miles or six times round the course. There were four entries. After the second round one of the condiates was so far in the rear, owing to baulking at a hurdle several times, that he gracefully retired. He had not much forsight or he would have kept on. In the fourth round one horse fell at a hurdle and, because the rider was slightly hurt, he retired. Early in the fifth lap one of the two remaining horses started to buck, threw his rider, then bolted for the bush, his rider after him. But he was seen no more. Lewis was still left in, but when a quarter around on his last lap a newspaper blew across the track in front of the horse. It bolted, his rider could not pull him up or control him. He went right through the town, and out of sight and, after an absence of about a quarter of an hour, returned to the course at the place from which he had bolted, completed the lap and won the race in a canter. Lewis was a rare boy, one of the aristocratic larrikins who are in for any devilment, and care for no one. And, of course, like the rest of the town, he got too much to drink. Beer has varied effects on different people, and makes them all kinds of nuisances. Some laugh, and can't help it, they would laugh at their grandmother's funeral. Others cry under the most jovial conditions. Some try to sing, (note the word try), these take the Belt for the Nuisance's Championship. Others want to fight. Lewis was of the latter temperament.

On this occasion he ran up against a big buck of navvy about sixteen stone, and as strong as a horse, but with absolutely no science. Lewis was about eleven stone, a splendid boxer and very agile.

The local constable and I watched the fight for about 20 minutes from a bedroom window. Lewis was getting in some splendid hits, but he may as well have been punching at an elephant while the navvy, disregarding his punishment as if he did not mind it, would occasionally get in a sledge hammer swing that knocked Lewis about two yards away. Lewis had no hope and, in spite of his science, was being gradually exhausted and demoralised by sheer brute strength. So the constable made his appearance and stopped the fight, threatening to run them both in if there was any more disturbance of the peace.

Harry Banfield's Hotel at Tarrawingee c.1892. Note the original Torrowangee spelling, taken from the original Government survey map.

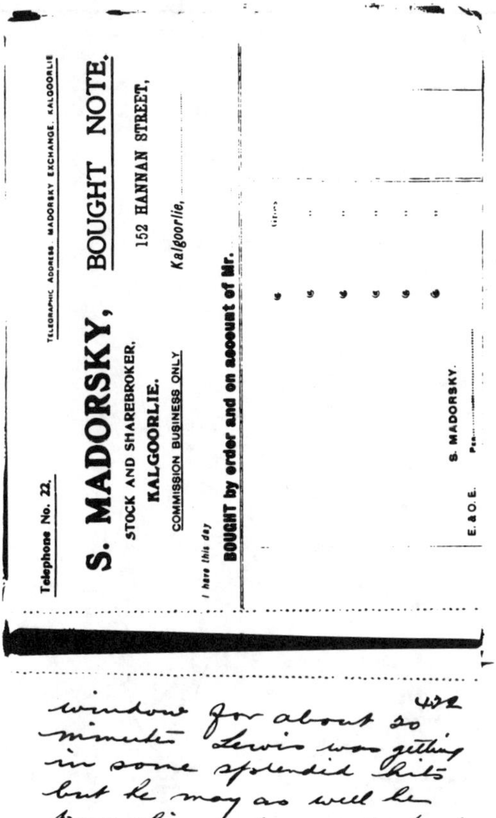

Telephone No. 22. TELEGRAPHIC ADDRESS, MADORSKY EXCHANGE, KALGOORLIE.

S. MADORSKY, BOUGHT NOTE.

STOCK AND SHAREBROKER, 152 HANNAN STREET,

KALGOORLIE.

COMMISSION BUSINESS ONLY Kalgoorlie,

I have this day

BOUGHT by order and on account of Mr.

S. MADORSKY.

Per

E. & O. E.

472

window for about 20 minutes Lewis was getting in some splendid hits but he may as well be punching at an elephant while the navvy disregarding his punishment as if he did not mind it would occasionally get in a sledge hammer swing that would knock Lewis about two yards away.

Lewis had no hope and in spite of his science was being gradually exhausted

W.G.M.'s original handwritten notes of the facing page number 60.
The entire notes for this book were similarly written on the reverse side of twelve broker "Bought Note" books.

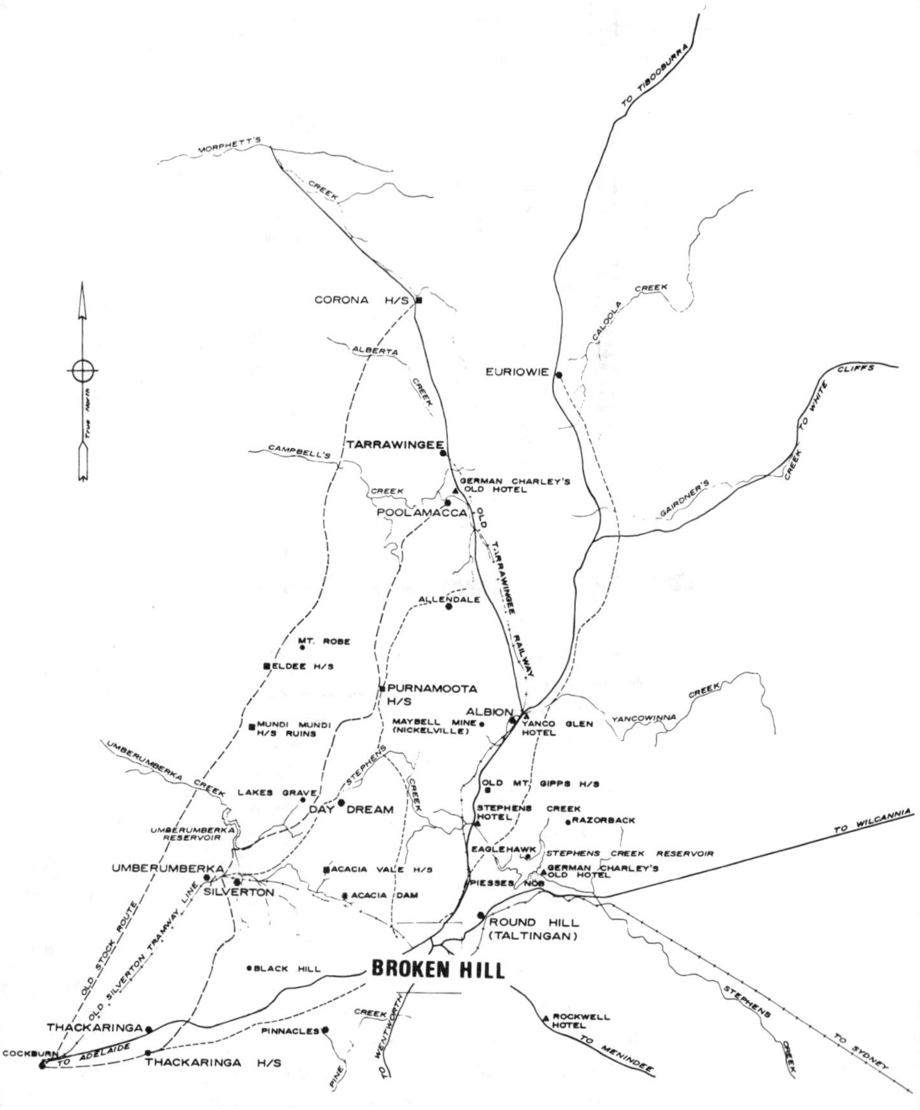

LOCALITY MAP
BROKEN HILL DISTRICT

12

Aesculapius (1893)

The Tarrawingee Coy. had a Medical Fund on similar lines to that of the Broken Hill Mines, to which all the workmen contributed by means of a deduction from their pay. The company in this case paid the doctor a salary and, of course, got the cheapest man available, irrespective of his genius and ability.

Dr. Johnson was the successful applicant, and I call him Doctor only out of respect for his honoured profession. He was a graduate of the Melbourne University, but did not have his Doctor's Degree. He was an M.B., which I understand means Medical Bachelor. That he was a bachelor we knew, because he had luckily escaped the dangers of several leap years. But how he attained the distinction of being a Medical Bachelor is probably a secret between him and a university professor. What he didn't know about medicine and healing the sick would make a splendid treatise on the Medical Science.

He was about twenty five years of age, and had all the confidence and cheerful optimism of a renowned physician. I suppose mathematics is not a subject required to obtain medical degrees or, if it is, then he must have been lucky. We once set him a problem to test his knowledge, and he very successfully demonstrated his lack of it. The problem was to multiply 7 miles, 5 furlongs, 47 yards, 2 feet 8 inches by 4 tons 14 cwt. 2 qrs. 17 lbs. He spent all his spare time for a week, and destroyed quite a bundle of otherwise useful writing paper in trying to solve this riddle. It was only when it dawned on his expanding intellect that he realized we were fooling him, and that the thing was impossible. We loved him for his charming innocence and christened him "Aesculapius" the God of Medicine.

We had two children by this time, a boy and a girl, and there was a rather significant prospect of this number being added to, so our beloved mother in law was again on the scene.

It was during this visit that Aesculapius was made happy and important by an epidemic of bronchitis among young children; about twenty were laid up at the same time so Aesculapius was very busy. Our little girl about two and a half years old, was one of the victims. The poor kiddie was lying in bed suffering with a pain in her chest, wheezing and gasping for breath, and was very feverish. I suggested we send for the M.B., but Mrs Brown said she had brought a book, the first series of Professor Kirk's papers on health (There are eleven series of these books, now published in one volume), and because there was a chapter on bronchitis, she proposed that we should try his instructions. These were to put a hot bran poultice on the back between the shoulders and very cold towels on the chest. The former would keep warm for forty or fifty minutes, but the latter were to be changed frequently, i.e. every two or three minutes.

After an hour or two of this treatment the congestion entirely left my daughter's chest, and she was breathing naturally. But her fever was extant. The instructions were then to place a few cold towels on the stomach and, in an hour or two, the temperature was reduced to normal. The child was up next day playing on the floor.

Of the twenty or more patients in town, two of them died. None of them recovered under three weeks, while some developed what the medical faculty call chronic bronchitis. Chronic used in this sense is a term expressing medical inefficiency, although it is not so expressed in the dictionary. Can we believe that these men, who should be like the great physician of old, do not know this simple remedy for so simple a complaint? If it be that they honestly do not know, then I am sorry for their system of education. One is almost tempted to believe that the wise among them must know, but would it be expedient to employ a remedy that any old woman could apply, that would cure in a few hours, when a nice cheque might be earned by allowing the patient to recover by natural recuperation, in spite of the useless medicine.

We know that the public are a bit to blame for this state of affairs, because we have known many people who would think their doctor was a fool and neglecting them unless he gave them a "bottle o summat". I believe doctors trade on this weakness which is so profitable to them. A little misfortune which happened to me some time later accentuated these opinions of mine, and convinced me that many of our maladies may be cured by simple means.

Our air compressor wanted a little attention to one of its valves and, in order to accomplish this, it was necessary to close a plug cock between the air receiver and the air compressor. There was a pressure of 80lbs. per square inch in the receiver. And the cock had not been used for some time and was clogged to some extent with carbonised oil. I had tried with wrench and hammer to close it, but could not move it. Then very, foolishly, I took the nut and washer right off the plug of the cock and hit it once with the hammer. It shot out like a ball from a cannon, fortunately away from me, but a fierce blast of compressed air and dirty carbonised oil struck me fair in the face and knocked me down. I was taken home and the doctor sent for. It was at first thought that my eyesight was destroyed because I could not open my eyes, but the doctor opened them, wiped the muck out of them with a silk handkerchief, and pronounced them unimpaired. My face was all pitted with carbon specks, but these were soon removed by rubbing my face with vaseline.

My eyes became very painful, excruciatingly so, and the doctor provided a lotion, I think sulphate of zinc, which increased my pain exceedingly.

This went on for several days. I was confined to a dark room, my visiting friends came in with sad faces and sympathised with me in my distress. My groans, which I could not suppress, so distressed my wife that she asked Aesculapius whether he could do something to relieve the dreadful pain. He regarded this as a sort of a joke, for he laughed out loud and said, "Oh! we can't do more than we are doing, he will be like this for six weeks yet." This was at mid-day. When he had gone I asked my wife to look in Kirk's book to see whether there was anything about the eyes.

Sure enough there was a chapter describing my accident almost exactly, except that it was with high pressure steam instead of air. My engineering knowledge informed me that high pressure steam does not scald when escaping, so the cases seemed exactly similar. Kirk explained that the eyes were not injured, but that the shock had stopped the nerve action which flowed from the top end of the spine. The stoppage of the nerve action caused congestion in the eyes, and the congestion caused the pain. The remedy was to start the nerve action by a hot bran poultice on the back of the neck and cold handkerchiefs placed over the eyes, changed as in the case for bronchitis.

The result was miraculous: in two hours I was as well as ever I was, the pain had gone, the heavy blinds were removed from the window, and I was like Mother Hubbard's dog when she brought him a coffin. That is, I was sitting up laughing. When Aesculapius came in at 5 o'clock he was so astonished, he could not believe his eyes. He asked us what we had done and, when we told him, he said it was a miracle, he had never heard of such a thing before.

Yet thousands are spent every day educating these scientific human healers. Yet, with all their alleged science and university education, they do not know this simple anatomical fact. My mind reverted to poor old Grant who spent hundreds of pounds on doctors and went to his grave a miserable man after suffering pain for years in spite of the best medical ability that could be obtained, when, had someone known of the fact I have written, any old woman could have cured him with three pennyworth of bran and a drop of God-given cold water.

I do not wish to be misunderstood, I do not condemn all doctors. I have met some real good and clever men, who do their best to heal and, in such cases as are not amenable to these simple remedies, and in surgical cases, they are a blessing to society. What I say is that their training is on a wrong hypothesis; it is founded on the fallacy that all ills of the human flesh are curable by means of drugs, especially poisonous drugs. Consequently, they are taught nothing about the remedies provided by nature, or the wonderful effects of heat and cold intelligently applied. Again, the system of hiring doctors is all wrong.

The present method is an inducement to keep people ill and not hurry the cure. An epidemic is a regular Godsend to them. They are in the same position as an undertaker. They thrive on the misfortunes of the people. The same thing may be said of surgeons. There is a great temptation to perform an operation to cure a bellyache that could be cured with a dose of castor oil. But the oil would cost only a penny, the doctor make only one visit, and charge 10/-, while an operation would enable him to charge fifty guineas. Some system of medical service should be inaugurated whereby it would pay the medical faculty better to keep people in good health, then they would come back to natural remedies, and the death rate would be marvellously reduced. Another thing is that a person who can cure these ills is liable to prosecution if he is not registered as a doctor. And he is not allowed to charge, so that one cannot do the amount of good one might do unless he has independent means and ample leisure.

Our friend Aescupalius was not a surgeon, he was just an M.B., as I have already explained. He nevertheless had his uses and, in the case of accidents, which were frequent and sometimes rather serious, he was good at rendering first aid.

We were using an explosive known as Rack-a-rock. At the time it had two constituents, a solid and a liquid, which, when kept separate, were quite harmless. But when combined they became a powerful explosive. We were boring deep holes and using big charges for the face of our quarry was sixty feet high, and we could blow down fully 100 tons with one shot. Some of the stones would be too big for our rockbreakers or for spalling and would have to have 'pops' put in them to break them up.

Premature explosions were fairly frequent; I have seen as many as four men killed at one time, their legs and arms broken, often hanging by a piece of skin. It required the trained callousness of our Aesculapius to handle these cases and fix them up to be sent into the hospital or the morgue, as the case required.

So, although I suspect he never cured anyone from the simple ailments that prevailed, he had his uses to help in case of accidents, and to supply amusement to us in his leisure moments.

THE TARRAWINGEE RAILWAY AND FLUX COMPANY
THE OPENING OF THE WORKS

Forty miles from Broken Hill, is the little Hamlet of Tarrawingee, brought into existence by one of the most important discoveries made on the Barrier. Seven years ago Broken Hill, now a city of 25,000 people, had no greater pretensions to the name of township. Will the demonstration which marked the opening of the Tarrawingee Railway and Flux Company last week be the datum-point from which in the future we shall record the history of another city of the plains?

Nowhere in Australia has there been carried out on so elaborate a scale a private fete similar to that held at its inauguration, and few public ones have been so successful or resulted in so much pleasure to those who attended it. The occasion was worthy of a demonstration, for next to the principal mines of Broken Hill, Tarrawingee is most important. The mines produce ore in thousands of tons per week, but without limestone no smelting could be done.

In 1889 Tarrawingee Hill was discovered to be a huge mass of splendid limestone, or, as the stone was crystalline in texture, marble. Leases were at once taken up, and others added, until within a short time the whole of what is now the property of the Tarrawingee company was held by a syndicate.

One great obstacle, namely, distance from the mines, appeared to loom largely against the prospects of the syndicate. Forty miles of country, not well watered, would have to be crossed by teams if the mines were to be supplied with flux, and it was doubtful whether any profit could be made out of the venture.

But the syndicate was composed of energetic men, and they determined to test the value of their find in a very practical manner.

The Tarrawingee syndicate employed experts to visit and report upon these, and to thoroughly prospect the country for miles round in order to make sure that no larger, better, or more conveniently situated deposit of limestone existed, and none was found. Meanwhile, a flying survey from the Hill to Tarrawingee was made, and it was seen that no difficulty would be met with in laying a railway line. These things being demonstrated to the satisfaction of the syndicate, they next sought the aid of capital. Happily for the success of the undertaking they laid the matter before Mr J.S. Reid, a gentleman possessed of great energy, and known far and wide as a man of honourable and upright dealings.

In July the company was floated and in November, just 12 months after the discovery of the Hill of marble, a contract to build the line was let to Messrs. Baxter and Saddler. The sleepers required were then growing in the forests owned by Messrs Miller Bros in West Australia and the rails had to be ordered from England.

This piece of railway making beats everything yet done in Australia, and reflects the greatest credit on the engineers.

There is no doubt that the laying down of the Tarrawingee line and the opening of the flux quarries have been a great undertaking for a private company, and not only is profit assured to the shareholders in the Broken Hill mines, they can feel certain that no stoppage or delay in the matter of supply to the furnaces will occur, and that by being given a pure article the smelting of the ores from the mines will be done better and more economically henceforward than ever before has been the case.

The Argus, Saturday, June 20, 1891

13

Mugs And Martyrs (1891 - 1894)
(The Birth of The Labor Party)

Mr Jack Brougham was the Manager of Poolamaca Station. When his men were not poisoning rabbits, they were attending to the sheep. He had an engineer looking after his pump and other machinery.

But sometimes the best engineer finds a problem that is too much for him, especially if he has no brain capacity or has not been advised in his youth "to think for yourself, lad". Mr Brougham was in great trouble, he had a large vegctable and flower garden at the homestead which he irrigated from a dam by means of a steam boiler and pump.

His engineer had been unsuccessful for over a week to get the pump to deliver water. And, being hot weather, the garden was languishing. The engineer was at the end of his resources, so Mr Brougham came in and asked me to go out and have a look into the difficulty.

Steam was up so I started the pump. It worked smoothly, but would not pump water. I examined the water plunger and the valves, they were in perfect order. There was only one thing that could be wrong and this was the suction pipe. You see, I adopted the same system as the doctors, that is, diagnosis by elimination.

I got a spanner and started to take off the suction pipe, to find that the bolts had never been screwed up. The flange was admitting air, so I screwed up the bolts and started the pump - up went the water into the tank. The whole process did not take more minutes than the mug engineer had spent days, and Mr Brougham smiled and looked sarcastic.

Then came the great strike. The leaders of the unions, variously described by the mine owners as "blatant agitators" had got, as Holly would say, "good and ready to put up another fight." So all the men were called out, including our crowd, and the fight went on merrily for some weeks.

Our directors sent some "gentlemen" up from Melbourne to take the place of the strikers. The unionists called them "blacklegs." This was too polite a name for them, they were the scum of Little Bourke Street. We had to lock our doors, a thing we had never done before. Most of them brought fishing tackle: they had been told it was a great country for sport, especially fishing, although the nearest fishing was the Murray River, some hundreds of miles away.

Big meetings were held daily in the reserve at Broken Hill. The speakers were rather on the brilliant side, but necessarily hostile to what they call the "capitalists". Sir Henry Parkes was Premier of the state at that time and, at the request of the mine owners, sent a small regiment of soldiers and a dozen or two policemen.

These were quite unnecessary because the meetings were all orderly. But it was evidently considered good politics to incite the mob to riot and thus have an excuse for shooting them down and smashing their organisation. But, so powerful were the leaders, and so great was their control over the men, that nothing happened the way the bosses wanted. Now and again some man with a surplus drink or two would create a disturbance and the police would seize the opportunity to create a scene. There would be an attempt at rescue, and talk about raiding the jail, but the strike leaders would rise to the occasion and quell the disturbance, advising the men to suffer arrest or any other injustice quietly, as they, the leaders, were quite prepared to do, and not work into the enemies' hands by creating a disturbance.

This was all very annoying, and things were not going as the Government planned. However, there was another chance: The leaders were speaking at times rather insultingly about the bosses and the Government. They would perhaps call Sir Henry Parkes 'a muddling old fool'. So he proceeded to prove that the accusation was quite justified, by sending a few detectives over to make reports of these speeches. They soon accumulated volumes of interesting literature, characterising the merits and demerits of the premier and minister. This was regarded as sedition and, by direction of Sir Henry, all leaders, numbering about a dozen, were arrested. As additional proof of Sir Henry's claim to the character given him by the leaders, and to give them a better chance of imprisonment, they were taken to an outback town called Deniliquin and tried by a jury of squatters. They were given sentences varying from three to six months.

Sir Henry thought he had won a great victory when, shortly after this, the men, being deprived of their leaders, gave in, and the strike collapsed. But the resentment caused by this despotism and a rather comprehensive black-list kept by the mines offices, whereby hundreds were reduced to starvation and distress, caused a great many business men and others, who were not supposed to be on the side of the men, to give them their sympathy and support.

Matters being at last all quiet, the Government let these men out of jail at the end of about three months, just about three months before a general election.

My father was travelling from Sydney to Melbourne and got into a carriage with about half a dozen men. He thoroughly enjoyed their company, they were witty and interesting. They discussed all sorts of subjects with intelligence and depth of knowledge. They were all gentlemanly in appearance and manners and Dad wondered who they were. He had not enjoyed the society of such men before. Imagine his surprise when arriving at Albury, a mob rushed the carriage and cheered his companions to the echo. He then learned that these were jailbirds, agitators, and sedition mongers, they were Messrs. Sleath, Cann, Ferguson, Thomas and Beaglehole, returning from jail. It took a genius like Sir Henry Parkes to discover the criminal instincts in these men and make martyrs of them. I would suggest they were mugs made martyrs.

W.J. Ferguson was a natural wit and the best man at repartee, perhaps, barring C.E. Jones, I ever heard. I do not remember any of his public sallies, but one of his retorts while in jail is well worth repeating.

A lady of sentimental and polical turn of mind had called to visit him and, after consoling with him in his incarceration, remarked, "Well, never mind Mr Ferguson, stone walls do not a prison make, nor iron bars a cage."

"Perhaps not," replied the prisoner, "but I must say they make a damned good substitute."

The general elections came along. At this time, the great questions of the day were Protection and Free Trade. Sir Henry Parkes and George Reid, later Sir George, were their champions respectively. That is to say, the great question of difference between these two great politicians was whether Sir Henry Parkes would pile on heavy customs tariffs and call them protective measures, or George Reid would impose the same duties and call them a revenue tariff.

As far as the people who had to pay were concerned, it did not matter a straw who was in power but, of course, they did not know this. It is a secret known only to the politicians and perhaps it is unkind of me to expose the fraud. There was no Labor Party in Parliament at that time, or if there was a member or two, they were in negligible quantity.

However, so great was the feeling raised by this dastardly act of despotism, that nearly every candidate was opposed by a man standing directly in the interests of labor.

There were five or six constituencies in the Silverfield district. Each of these put up what they called a liberal candidate, and he was opposed by one of the mugs made martyrs.

Suttor and Ferguson were the candidates for the Tarrawingee district. It was well known that Suttor's expenses were being paid by what was called the Liberal League. I think they had difficulty in getting anyone to stand at all, and Suttor was not likely to set the town on fire with enthusiasm.

I nearly missed the number of my mess by meeting and talking to Ferguson at the pub when he came out to speak. The meeting was quite accidental and, because I knew him on Block 14 where he had been engine driving, we talked for some time and had one or two drinks. A few days later I received a letter from a particular friend of mine who was a member of the Liberal League, warning me that I had been reported for "hob-nobbing", whatever that may be, with Ferguson, and to be very careful or I might lose my job.

To test my loyalty, the manager asked me to propose a vote of confidence in Mr Suttor when he addressed the constituents. Of course I would, and of course I did, and this seemed to satisfy the simple minded League for I heard no more about the matter.

That they were simple-minded optimists was illustrated by the management sending the storeman around the quarries with a memo book to ask every man who he was going to vote for. My department was the venue of his first attack, and I asked him what he was doing. When he told me, I asked him if he believed he would get trustworthy answers from the men with a black-list and probable sack awaiting them. In his cheerful optimism he thought he could rely on their words. When he returned I asked him how he got on. "Oh splendidly. It's a sure thing for Suttor. I have 300 names for him and less than fifty for Ferguson". I was surprised to find there were even fifty brave enough or fool enough to risk the sack. But I offered to bet him five pounds to one that his figures would be reversed at the election. He was not betting.

When the results were known, every Broken Hill electorate had returned Labor members with from three to one or five to one majorities. Ferguson's majority was nearly five to one, and the figures for Tarrawingee were just about 300 to fifty in favour of Ferguson. That the political despotism had affected the whole state was shown when thirty two Labor members were elected in a government of seventy members. With so greatly reduced numbers in the two government parties the Labor party held the balance of power and could control the government. To avoid this anomalous position, Parkes and Reid agreed to form a coalition government, calling themselves the Liberal Party. The only question to settle was whether they would call the customs taxes a protectionist tariff or a revenue tariff. Thus was the first great Labor Party in Australia orginated and fostered by despotism and foolishness.

The Labor Party gained nothing for several years. I think they even lost a few seats in the course of time until another genius of the name of Wade waded in. Sir Henry Parkes was dead and Wade became Premier. As I said in my introduction, some folk do not learn from precept or example but must gain their own experience.

There was a strike at Newcastle, N.S.W., and Wade did practically the same thing as Parkes had done in Broken Hill - put all the leaders in jail. These in turn, became mugs made martyrs and, at the next election, the Labor Party swept the polls, returning with a good majority. So was born the first Labor Government in Australia. I do not say but that the Labor Party would have ultimately become as powerful as it is today, but such short-sighted men as Parkes and Wade raised them to power at least twenty years earlier than would have been the case by natural evolution.

"The Leader" 3 September 1892

Much More Wealth In Broken Hill

NINETY-FOUR-YEAR-OLD PROSPECTOR'S OPTIMISM

When Mighty Mines Sprang Out Of Mulga Scrub

PIONEER of many mining fields in Australia, including the Golden Mile, Coolgardie, and other famous Western Australian areas, Arltunga, Tennant's Creek, and White Range in Central Australia, Lamb prospected over the Broken Hill country in 1880, when the Hill was beginning to emerge from mulga scrub into one of the biggest mining towns in the world.

"Jimmie," as he is known to a wide circle of friends, celebrated his 94th birthday recently, but his recollections of those early days at the Hill are very clear.

Lamb says that the Broken Hill line of country was first prospected by the late Mr. Patrick Green, a once well-known Menindie storekeeper in the sixties of last century.

Prospects were not promising enough and the work was abandoned. The men were then shifted to a show three miles north of the Mt. Gipps Hotel, known as the Yellowstone, and owned by Messrs. S. Brown and R. B. Pell.

There a 50ft shaft was put down for copper. That claim was also abandoned, and nothing was done with Broken Hill until about 1874. Then Mr Charles Nicholls, the original prospector for silver at the Barrier, went over from Thackaringa and pegged out the famous No. 3 Block, situated between the Sultan and the Round Hill Companies' leases.

After Block 3 had been pegged out, said Mr. Lamb, the late Charlie Rasp pegged out the original seven blocks of the Broken Hill Pty., thinking at first he had found a mountain of tin.

Then Block 17 was pegged out by Mr. Otto Fischer, and was named by him the Cosmopolitan. Afterwards it became the property of the Broken Hill North Company. Several weeks later Blocks 5, 6, and 7 of the Broken Hill South were leased by Mr. William Maiden, of Menindie.

Block 7 of the same company was secured by Mr. White, also of Menindie, and Mr. Thomas Nutt pegged out Block 9, which he named the Elizabeth. Afterwards this was the site of the Broken Hill Central.

Shortly after the Broken Hill Junction—then known as the Great Northern Junction—was leased by Messrs. Penglase and Carson. "The first I heard of Broken Hill," Mr. Lamb relates, "was when we were camped at Pernamoota, then known as the Soakage.

First Sight of "Hill"

"The township of Pernamoota had just been surveyed. People were rushing out in hundreds from Silverton to secure allotments where the Lubra mine had just been floated—also that wonder of wonders, 'Morris's Blow,' found by Morris and the Nolan brothers.

"Those who thought anything about Broken Hill in those days called it a 'mountain of mullock' that would never pay. I was camped with a man who owned three one-fourteenth shares in it. He would have gladly given the lot for a fiftieth in the Lubra or 'Morris's Blow,' or the old Pluck-Up mines.

"The first time I saw Broken Hill was when I was sent out with others to make a tunnel in it in 1874. The owners intended to put a tunnel through the Hill near Rasp's shaft, but the idea fell through and the sinking of a shaft was decided upon. A contract to sink 150ft was let at £2/5/- a foot.

"The shaft had already been sunk 65ft in carbonate of lead by a man named Rosewarne. At that depth Daniel and Jack McKay, Sandy Kemp, and J. Wiles started, and soon sank a shaft to the 100ft level.

Worth 1½ Millions

"By that time the lode went out of the shaft and the remaining 50ft was sunk in country.

"The highest assay they could get gave only 16oz. of silver to the ton, but there was generally a good return for lead. No wonder the owners became disheartened, especially as at that time wonderful ore was being raised all over the Pernamoota and Day Dream districts, and the newly-discovered Mt. Gipps mine which gave their tens of thousands of ounces to the ton."

Then Phillip Charley, a boundary rider, who recently died in Sydney at the age of 74, when turning over some ore, came across a bit of chloride, and, on making a closer search, found several pieces of carbonate of lead containing chlorides.

Rushing over to Lamb's camp he cried, "Look at that, boys! I'm in it at last!" He threw the stone on the table.

"Only a bit of carbonate of lead," said someone.

"Carbonate of lead be ——," cried Charley. "Don't you see that?" and he pointed to the specks of chloride.

"It made us laugh to hear him talk so extravagantly," Mr. Lamb said, the other day. "But little did we know that three years later those shares would have a market value of nearly 1½ million pounds."

Men Of the Day

With the discovery of chlorides came the real boom in Broken Hill. Soon afterwards, Mr. Lamb recalls a meeting of owners was held, and Mr. William Jamieson was appointed manager at £500 a year, and Mr. Sleep, mine manager, while a little later Mr. Fawcett was appointed assayer.

There was not a house of any kind in Broken Hill then. Chlorides were found in many other spots, and things became lively.

"We drove into Silverton next morning with a buggy-load of specimens, and the sight of them drove the Silverton people wild about Broken Hill. A day or two later," Mr. Lamb went on,

"Mr. John Jones, a boardinghouse-keeper from Silverton, drove up to our camp and, by a bribe of a dozen eggs — a great thing in those days —obtained the sole right to direct dining-rooms on the syndicate's leases.

"This was the first business started in Broken Hill. But one had to be careful in those days. We found that nine of the dozen eggs were rotten, while the others had a most ancient aroma.

"A swindle?" said Mr. Lamb. "Yes, it was. But I have often thought that those eggs were but the forerunners of other still more rotten swindles.

"About this time, Messrs. Logan and Downie, who had been prospecting near the Maybell, pegged out the leases known as the Diamond Drill, for the Brisbane Block Company.

"Mr. Logan had a bit of hard luck before this. It appeared that he had found chlorides in Rasp's shaft a few days before Phillip Charley, and he had written to Brisbane for money to buy in. But as there were no telegraph wires or trains in Broken Hill at that time, several weeks elapsed before he got his money. By that time the cat was out of the bag.

"Logan's partner, Downie, had £900 in the bank at Silverton, but he would not believe that Logan had found chlorides. So he was out of it, too.

"Shortly after this the Victoria Cross was bought by Messrs. S. Brown and Morgan D'Arcy from Daniel McIntyre and Don O'Connor. A few days later the South mine was bought from William Maiden, of Menindie, for £1000 by Messrs. Jamieson and Keats.

"The latter beat my old friend, Charlie Chapple, by only a few hours in their race to Menindie to buy the mine which was soon to be worth 1½ millions.

"Logan and Downie determined to trace, if possible, the Broken Hill lode past the alluvial flats south of the South Broken Hill Company's leases. They succeeded in finding the White Lead.

"People were then under the impression that the Pinnacles was the continuation of the Broken Hill lode. I once pegged out two blocks on the White Lead. One of them was afterwards Central White Lead, and the other Block 8."

Then, recalls Mr. Lamb, came a man named Delamore, who was received with astonishment when he announced that he intended to build an hotel. He built it in the scrub, and did a roaring business.

Mr. Lamb does not expect those great old days to come again at Broken Hill, but he is firmly of the opinion that much more remains to be discovered in that wonderful hill of silver and lead.

BROKEN HILL is in the news again. A number of companies, backed by solid men in Adelaide and the Hill itself, have been formed in recent months to exploit areas surrounding the great mines there.

Jimmie Lamb, Australia's oldest prospector, believes that there is much hidden wealth in the comparatively untried country. And he should have a good idea.

FEDERAL

AMALGAMATED MINING MANAGERS' ASSOCIATION OF AUSTRALASIA.

CENTRAL OFFICE: MINING EXCHANGE, LYDIARD ST BALLARAT. JOHN SHARPE CENTRAL SECRETARY.

This is to Certify that

Mr. W. G. Manners *has been a*

Contributing Member

of the Ballarat Branch of the

F.A.M.M.A. *from* January 1898

[illegible] Emery *President.*

[illegible] *Vice-President.*

J. Sharpe *Secretary.*

Dated at Ballarat the 18th *day of* June *1902*

14

MacDonnell Ranges and James Lamb "The Prince of Prospectors" (1892)

A Bricklayer on the works, named James Lamb[1], had a letter from a friend advising him that his brother, who had been living in the MacDonnell Ranges, was dead, and that he had left to him some wonderful Mica Mines.

We formed a Syndicate and sent Lamb up to locate these Mines and, if they were as good as reported, to fix up the transfers.

Lamb went up to Arltunga but cound learn nothing about the location of these mines. He got in touch with two prospectors named Luce and Thompson who were in possession of a gold mine. He obtained an option from them and brought down half a ton of quartz which was sent to the School of Mines at Ballarat for tests and treatment. The result was a button of gold valued at approximately £10. This looked very promising and, in order to confirm the estimate of the mine, I obtained a month's holiday and was sent up to report on it. This would be about the end of the year 1892.

The journey by train, first to Broken Hill, thence to Adelaide, thence to Quorn, and on to Oodnadatta, occupied about five days and was monotonous in the extreme. The only interesting features of this part of the journey were the Coward and Hergott Springs. At these places were artesian bores at which a column of water shot out of the bores to a considerable height then fell like a large fountain. Small fish about the size of Minnows came up these bores and could be seen swimming about in the stream running away from them. They were quite blind, bearing testimony to the utter darkness of the underground river from which they came.

It was thought at this time that this underground river had its origin in the great dry lakes, Lake Eyre and others into which the waters from the tropical rains disappeared, and that the river emptied itself in the Great Australian Bight.

From Oodnadatta, it was necessary to take a coach for about 200 miles through sandy and stony country, the gravel for many miles on this road being water-worn and smooth, from which one would infer that most of the country had been under water at one time.

All this country was taken up as cattle stations and, because we called at several of these homesteads en route, it was necessary to take very little food with us. It was always salt beef we had at these places, and they always gave us enough to see us on to the next station. However, we came to one place where there was a sly grog shanty. Here we were given beef, and tea, and roast mutton, which was a treat after some days on salt beef.

On starting from this place we saw a flock of about 1000 goats. I asked what they did with this herd, and was informed that they were the source of mutton we had for dinner.

The journey to Alice Springs was uneventful and, from there, I had to hire a saddle horse and ride out with the mailman a distance of seventy miles to Arltunga, locally known as Paddy's Hole. I had not been on horseback for some years, and the affinity between a man and a horse is an acquired art. As we rode about forty miles before camping that night, I did not regard equestrian exercise as a pleasure.

The country was very rough and very beautiful and, for many miles, we travelled along creek beds that seemed as if the hills had been split and opened, then filled up a certain distance with sand and gravel. Sometimes the walls of the cutting went up almost vertically for approximately 100 feet.

We arrived at Arltunga on Christmas Eve to find a population of about twenty men and as many niggers of both sexes who worked as servants. They were in great distress because their Christmas cheer had not arrived. Three months previously they had ordered three barrels of beer and a supply of butter, bacon, raisins, currants, and other luxuries and necessities in order to carry out the festivities of Christmas in a fitting manner. But it transpired that the teamster who was entrusted with this very valuable cargo could not resist the beer, so tapped one barrel after another. In the three months spent on the road, which could easily have been covered in a month, he had drunk all the beer and eaten a large quantity of the provisions then, as a matter of fact, arrived about a fortnight after Christmas with three empty casks and a greatly reduced cargo of food.

We were thus reduced to a fare of damper and tea, but there being one tin of jam left, they generously broached it in my honor, but would not use it themselves, and evidently meant me to have it all. Of course, I was cracking hardy as the saying is, and it was pathetic to hear them pleading with me to "have some jam, Oh please, do have some jam," and they would shove the tin over to me. I really had to take some, otherwise I am sure they would have been unhappy.

If you want to find the true spirit of comradeship and generous heartedness, get right out into the back blocks. There you will find the tough, rough, old pioneers, who are afraid of nothing, but will share their last damper with a stranger, knowing at the time that they may themselves have to go hungry for days in consequence.

The next day I went out with Luce and Thompson to see the alleged mine, but was greatly disappointed to find there was absolutely no work done. And, with the exception of the small excavation from which the sample half ton was taken, there was not even a small costean (trench). Gold was visible in the outcrop which covered a good deal of the surface, but there was nothing to indicate that it went down any distance, so I could not conscientiously recommend it as a mine.

The next day, therefore, we started back for Alice Springs. The only event of any note on our ride back was a thunderstorm which occurred while we took shelter at a homestead. A flash of lightning passed through the blacksmith shop, killing a nigger who was holding a horse to be shod. The nigger was scorched from his head right down to his feet.

On arriving at Alice Springs in the evening I had tea then asked the way to the telegraph office. This was one of the Transcontinental offices from Adelaide to Port Darwin, about a mile out of the town. It was quite dark, but I was instructed to walk along the creek bed until I came to the P.O. lights. This I did and found the office all right. It was a large stone building which I had to walk around it a couple of times before I found the office, a small room in an obscure corner. I sent my telegrams then returned to the creek and, as I thought, started back home. I walked about a mile then came to a fork in the creek. This rather astonished me because I had not seen it before. However, I walked along one branch for some distance and, feeling I should have reached town, returned and walked along the other branch. I had not gone far before I saw lights in the distance and felt sure I was right. So I left the creek and made across very rough country for the lights. I did not know what the natives were like in these parts, and was a little scared. I met two niggers coming along, one being lame. I asked, "Which way Alice Springs?"

I was relieved to find they could talk English, and were probably not dangerous. The reply was, "Big Fellow, long way."

Then I said, "You come along show me."

Then Limpy replied, "Nothing walk, sore leg."

"You come along show me Alice Springs." I repeated.

"Come along." said Limpy. I followed them for about 200 yards and came to a large clearing in which were about 100 natives, all men, all stripped naked and painted and decorated with feathers and boughs, evidently holding a council prior to a corroboree.

When Limpy went up to the chief and told him about me, a great laugh went up from the crowd. I could just imagine the feeling. "This plurry fool white man lose himself." An argument went on for some time then a big fellow, fully six feet, stood up and came towards me, Limpy following.

"Which way walkem?" he asked.

"Alice Springs." I replied.

"Come along." said he and led off. Limpy, who could not walk with 'sore leg', coming too, and, after about a mile and a half, I found myself at the Post Office again. I had walked up the creek instead of down. An interview with the Postmaster raised a smile, but he very kindly saddled a couple of horses, and we got safely back to Alice Springs.

The next morning, when the coach arrived at the hotel, I noticed that the driver who had brought me up was not there, but another man was taking his place because the original driver desired to stay for the races were being held the next day.

The new driver was slightly intoxicated. The publican presented him with a bottle of whiskey and we started. By frequent applications to the bottle, my driver was soon beyond control, slashing at the horses who were going full gallop. The only means of preventing an accident would have been to knock him on the head to put him to sleep, and drive myself, but this idea did not occur to me. There had been rain in the ranges and we soon came to a creek about fifty yards wide with about a foot of water in it. We dashed though this and came safely out on the other bank, having kept on the road by some miracle. About a couple of hundred yards further on we came to a similar but wider, if anything, creek. We dashed into this, the whip flying, the horses excited, and we came out on the other side of the road into a bog up to the axles.

The driver was too drunk to be of any use so I took the horses of harness, out and hobbled them and turned them out. There was a saddle in the coach so I took one of the horses and rode back twenty miles to Alice Springs to get the original driver. This was a loss of a day and a half. And, if I did not get to Oodnadatta on a certain day, I would have to wait there a month.

I got the driver and a couple of fresh horses and we rode out early next morning. We found the driver partly sober and the coach unloaded. So we got the horses in, dug the coach out and were soon on our journey again. Unfortunately, there was a case of rum on board, consigned to someone on the road, and they broached this. So, it was not long before they were both drunk.

We camped in the bend of a creek for dinner. But being a day longer on the road than was provided for, this consisted of ''Johnnies on the coals'', that is, flour and water made into a dough and baked on the hot ashes.

Both drivers went to sleep. Because there was a little water in the creek, and some good pools, I went down to bathe. I was not long in the water when I heard a roaring noise and, looking up the creek, saw a great bank of water with logs and debris rushing down on me. I just had time to rescue my clothes when down it came over the pool. There had been a thunderstorm away in the ranges.

I woke the drivers and we put the horses in, but when we came to the creek we had to cross; it was about five feet deep and about 100 yards wide, with great logs and other timber sailing down. There was no chance of crossing, so we had to wait till evening. Another half day lost, and the train would not wait. However, about 5 o'clock, the water was only about three feet deep. But, as neither of the drivers could swim, or were afraid of the water, and as the nigger who minded the change of horses would not go in the water, I stripped off and waded through in front of the coach. By this means we landed on the other side safely.

Nothing further of a sensational nature happened, except that we got wet several times during the night and had to hang our blankets on the back of the coach next day to dry them. We caught the train, and arrived back in Tarrawingee within the month. But that was the end of the Mine. I expect it was worked about ten years later when there was a rush to this district, but it did not develop into anything permanent. However, I would not be surprised to see the MacDonnell Ranges a large mining centre when the railway is taken through the district.

1. Refer to Appendix III, containing correspondence directed to C.B.M. by James Lamb, then (in 1937) ninety five years of age. James Lamb was named by Smiler Hales as "The Prince of Prospectors".

   Lamb, in his letters to C.B.M. from 1935 to 1938, talks of the position of The Northern Territory venture from the vendor's point of view, and about early mining and W.G.M.

   James Lamb started prospecting in and around Silverton three years before Broken Hill was found.

   On arrival Silverton's population was fourteen (including James Lamb), - when he left the population was 40,000.

   First came to Broken Hill when population was seventeen (inlcuding James Lamb), increased to 21,000 by the time he left.

30-12-35.

From One Old Colonist to Another

A HAPPY STUDY taken at Glenelg Town Hall today of Mrs. B. Currie all smiles as she offers an ice cream to Mr. James Lamb. Mrs. Currie will be 99 next birthday while Mr. Lamb is 94.

Centenarian Attends Pioneer's 95th Birthday

August 1937

MRS. CURRY (left) and Mr. Lamb.

Mr. J. Lamb, of North Walkerville, who was born in South Australia, celebrated his 95th birthday yesterday. Among those who were invited to his home was Mrs. B. J. Curry, 100-year-old pioneer, whom Mr. Lamb has known since his schooldays.

About 10 people sat down to lunch at Mr. Lamb's home, where the table was decorated with a birthday cake adorned with 95 silver cachous, and with the number 95 outlined with more cachous.

During the day Mr. Lamb received a number of visitors and congratulations; by 1 p.m. he declared that he was enjoying his birthday much more than he anticipated.

Mr. Lamb, who was the first scholar to attend the Pulteney Grammar School in 1847, received an old scholars' badge and also a life membership of the association as a birthday remembrance. He was wearing the badge when his friends called to see him.

Mrs. Curry wore her coronation medal, of which she is justly proud.

Roseberry. North East Road south Australia
18th Nov 1935.

W G Manners. Esq. Kalgoorlie.

Dear Sir,

Seeing your addres in The Western Mail. I am Talking the liberty of penning to you a few lines. To Know from you. Are you a Relation, of my dear old Friend. a Mining Engineer. of Broken Hill. Some 50. or 55 years ago. He was also at Jarrawinga, If you are a son of his. Kindly wright me a few lines To Say. Yes- or No If you are a Son Il wright to you a long letter. I Have been prospecting since 1852. all over Australia. I am 93 years old last August. When I get a few lines from you. Il, reply phraps with a long letter to you. I Hope you are, Mrs Manners. will Remember me. Hoping to get a letter from you. Yours Truly

James Lamb.
Rosebeary. North East Road.
South Australia,

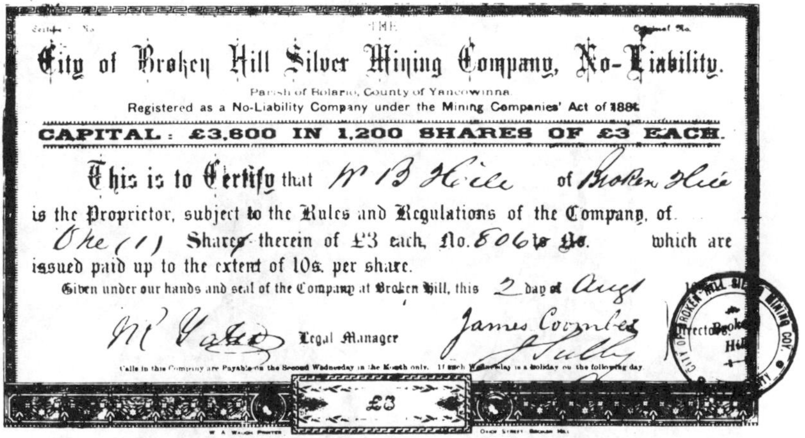

THE

City of Broken Hill Silver Mining Company, No-Liability.

Parish of Bolaro, County of Yancowinna.

Registered as a No-Liability Company under the Mining Companies' Act of 1886

CAPITAL: £3,600 IN 1,200 SHARES OF £3 EACH.

This is to Certify that W B Hill of Broken Hill is the Proprietor, subject to the Rules and Regulations of the Company, of One (1) Shares therein of £3 each, No. 806 to No. which are issued paid up to the extent of 10s. per share.

Given under our hands and seal of the Company at Broken Hill, this 2 day of Augt

Mc Vatt Legal Manager

James Coombes Directors

J Sully

Calls in this Company are Payable on the Second Wednesday in the Month only. If such Wednesday is a holiday on the following day

£3

W. A. Walsh Printer Oxide Street Broken Hill

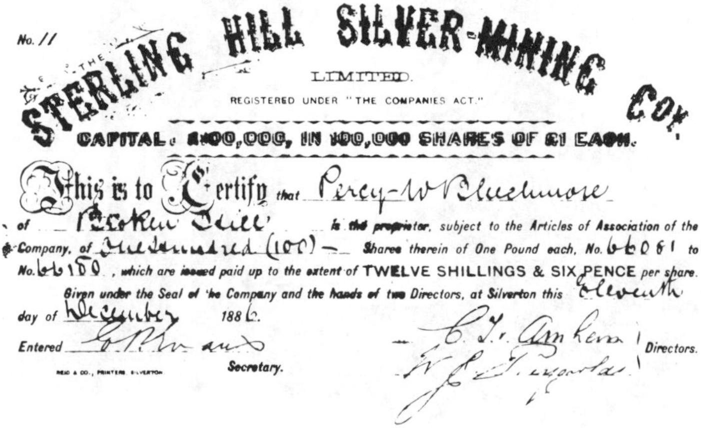

No. 11

THE STERLING HILL SILVER-MINING COY. LIMITED.

REGISTERED UNDER "THE COMPANIES ACT."

CAPITAL: £100,000, IN 100,000 SHARES OF £1 EACH.

This is to Certify that Percy W Bluchmore of Broken Hill is the proprietor, subject to the Articles of Association of the Company, of One hundred (100) Shares therein of One Pound each, No. 66081 to No. 66180, which are issued paid up to the extent of TWELVE SHILLINGS & SIX PENCE per share.

Given under the Seal of the Company and the hands of two Directors, at Silverton this Eleventh day of December 1886.

Entered G Rhoans Secretary.

C. T. Amherst Directors.

W P T Pogorlas

Reid & Co., Printers Silverton

The original 1892 "peg" marking Bayley's Reward claim at Coolgardie. The tree has since perished and been replaced by a concrete obelisk. This claim lead to the discovery of Kalgoorlie's Golden Mile, in the following year.

15

Groperland (Western Australia) (1895)

In the early nineties, Coolgardie in Western Australia was being talked about. Bayley and Ford had discovered the great Bayley's Reward Mine which was immensely rich, and then Kalgoorlie was discovered by Paddy Hannan, who pegged out Hannan's Reward. A big rush took place and Pearce-Brookman pegged out what afterwards turned out to be the Golden Mile.

Mr Martin of Messrs. Martin & Co. of Gawler, South Australia, who were supplying a large number of plants for the companies controlled by Mr Chas. Kauffman was asked to recommend an engineer. They suggested me. Thus I was placed in communication with Mr Kauffman and appointed to go to Western Australia.

So I sold up my furniture, took my family to Ballarat, then set sail for the Golden West. I embarked on a vessel of about 2000 tons called *Barcoo*. There were about 200 passengers all bound for the Goldfields.

I discovered on this trip that I was a good sailor; it turned out to be the roughest voyage I ever experienced. I had made two or three trips previously between Adelaide and Melbourne, and I have travelled thousands of miles since by sea, but have never seen anything anywhere approaching that storm coming round Cape Leeuwin. I was one of four out of the 200 who were not sea sick. I stayed up until about 11 o'clock at night watching the mountainous waves as the boat rose on them then sank into great valleys of water. The sight was simply magnificent, grand beyond description, and I thoroughly enjoyed watching the ocean's wonders.

I was under the impression that I had acquired my "sea legs", and was making my way to the companion-way with a view to bunk, when I noticed a lady lying on a seat with a rug over her. "Aren't you going down tonight?" I asked.

"I would like to," she replied, "but I am afraid to move." I was very gallant, "May I help you?"

"If you would, I would be so obliged."

So I assisted her to rise, took her rugs on one arm, took her arm with my other. We took about three steps when the boat made an extra special roll. Her feet skidded, hit mine, knocked them from under me, and down we came, I on top and, of course, having something soft to fall on. She was not so lucky. However, I helped her up and got her to her cabin without further mishap.

On arrival at Fremantle all was bustle and excitement, loading trains and booking passages for the Goldfields. The Sandgropers were inclined to resent this intrusion of the 'Tothersiders', as we were called. They felt that we were invading their territory and, in effect, taking it from them. They had been so long living under parochial conditions, and regarded the whole of Western Australia as their little bit of soil, that a rush of newcomers was disturbing their peaceful sleep. They were not used to the rush and progressive methods of the Tothersiders. The great majority did not know the use of money prior to this rush. For any service rendered, they would not take cash, but ask for an order on

John Throssell or some other of the twelve families that had, for so many years, controlled the destinies of this land. They had wooden ploughs, their waggons weighed about six tons, and would be loaded up with a ton of goods and drawn by six horses. When the Tothersiders landed with their waggons that weighed a ton, carried six ton, and their six horses simply walked past the slow moving groper teams, their resentment increased, and fights would result.

However, one Sandgroper stood out and showed that he was a man and a statesman, not only standing out above all his countrymen, but a statesman equal to any known to history. He was John Forrest, later, Sir John. He had risen to the occasion and recognised that the discovery of the Eastern Goldfields was to be a great blessing to his country which would raise it from the sleepy hollow that it was to a nation among nations.

He was rushing a railway through to the Goldfields with all possible expedition. He later, formulated and carried out the greatest water scheme the world has today,*(Note, written in 1924) to supply this great waterless area with that most essential commodity.

We booked to Southern Cross by government train. It was crowded and far from comfortable, there being none of the luxuries of travel we enjoy today.

There were several mines working at Southern Cross, including the great Fraser's Mine, which was principally noted for murders and burials in cement hearth stones of bodies of the victims of the infamous Deeming.

After inspecting the Mine and the camp of the murderer and the hearth stone, we booked by contractors' train to Woolgangie.

The Government train was bad enough, but this was much worse. But, when one is pioneering in the Wild West, one does not complain at little discomforts of this sort.

Steam boilers and cases of machinery could be seen lying along the track. These had been brought to the end of the line as the railway progressed then been thrown off the trucks for the teamsters to carry on to the mines. In a great many cases the ownership of this machinery was not located and it was never claimed. What ultimately became of it, I do not know.

On arrival at the end of the line, we still had a long distance to go by coach and, because the roads were badly cut up with the heavily laden teams constantly passing over them, the journey was rough and accompanied by a cloud of dust that obscured the view and almost choked the poor long-suffering passengers. The latter disability was not much loss; the landscape was mostly mulga and salmon gum plains.

There was no fresh water on any part of the Eastern Goldfields so at intervals along the road, were condensers built on a salt water well. The condensed water was sold at 2/6 per gallon. These condensers consisted of 400 gallon square iron malt tanks built in with stones and clay forming a fireside and chimney. The tanks were built with their corners vertical and horizontal, were filled with water by means of a bent pipe fitted into the tank below water level, and a cup or funnel at the other end above water level. From these malt tanks, large galvanised iron pipes led to large galvanised corrugated iron tanks supported on sapling stands. The vapour from the boilers passed into these cooler tanks, was condensed by the atmosphere, then drained into receiving tanks. One of these malt tanks would evaporate about 400 gallons of water in twenty four hours. It required cleaning out once or twice a week according to the density of the water treated.

On arriving at Coolgardie, which was at that time the great centre of attraction, we saw much activity and brick and stone buildings, business places and hotels being erected in all directions.

I went, with the Accountant of the London & Globe Gold Corporation, by whom I was employed, to Edward's Hotel. I found the appointments here quite up-to-date and almost equal to the hotels in older cities.

I could not see Mr Kauffman that day because he was busy playing poker with the Governor and staff who were visiting the Goldfields. But the next day I saw him and was sent up to Kanowna to erect a plant on the Golden Crown Mine. This was another coach trip of thirty six miles of heat and dust. On arriving at Kalgoorlie, I found the town in an embryo state. The Exchange was the only hotel erected at that time, a wood and iron building. All the other buildings were either wood and iron or wood and hessian, mostly the latter.

We stayed here only for a meal then coached on to Kanowna and took up my residence on the Golden Crown Mine.

Note: Appendix II contains a collection of W.G.M.'s poems written on board ship or on early arrival in the West.

THE THOUGHT THAT STRIKES THE ROVER (1896)

There's a thought that strikes the rover
Who's been tramping through the bush,
When the rush for gold is over
And he mingles with the push.
Who have been around prospecting
And have sold their little show
And are blowing it expecting
They will have another go,
When the boozing and the ladies
Have sent all their cash to Hades.

And the thought that strikes the rover,
Is that pluck and toil and graft
Has its little day of clover,
When it drank and sang and laughed.
Then he takes a job as trucker,
In his wallet - not a cent.
So he has to earn his tucker
And assist to pay the rent
Of the Landlord and the ladies
Who send all his cash to Hades.

Two views of Kalgoorlie's Hannan Street (looking east) at the time of W.G.M.'s arrival in 1895.
(Note 1st generation Exchange Hotel).

Exchange Hotel, 2nd generation c.1898 and 3rd generation c.1901.

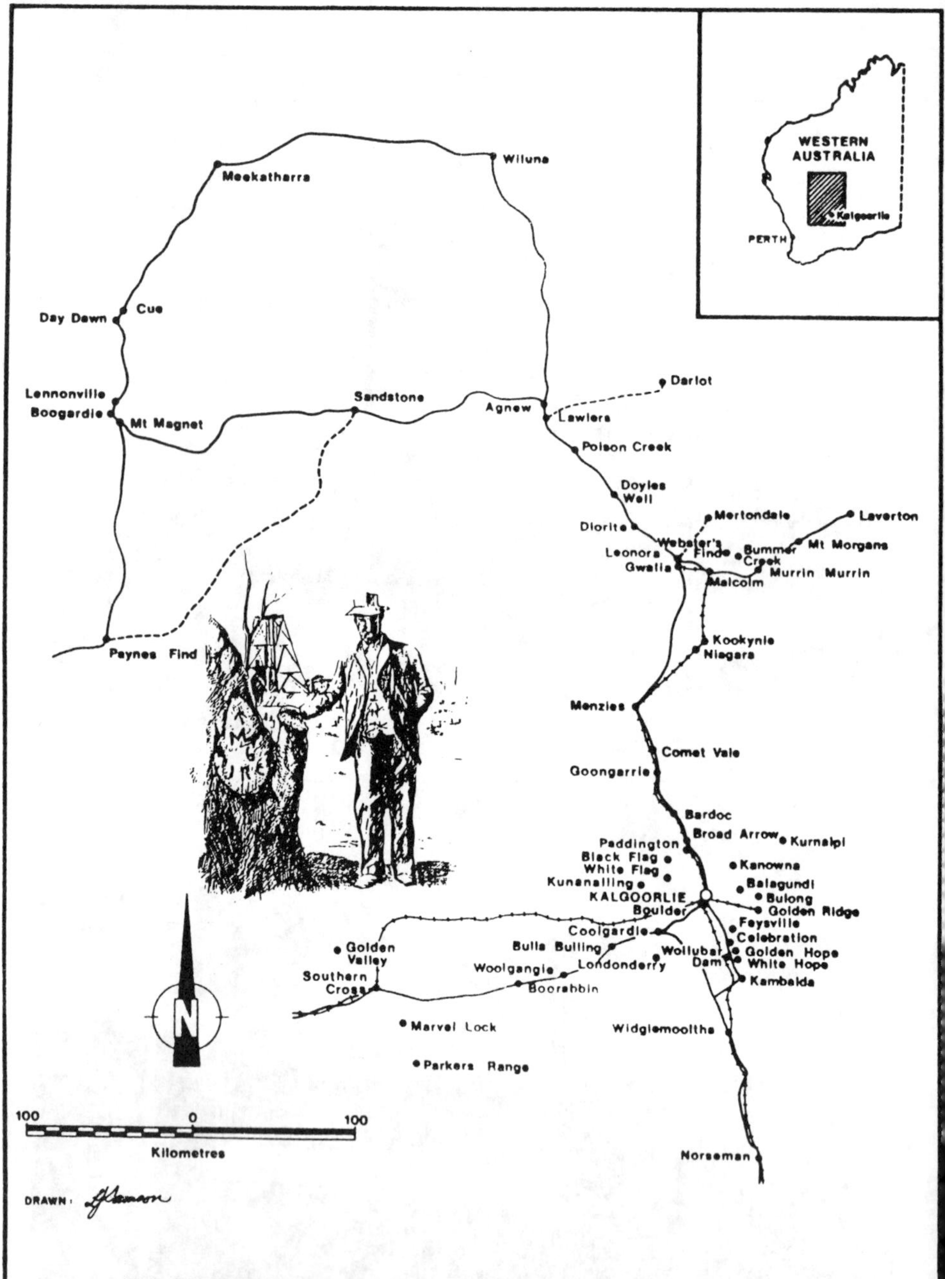

EASTERN GOLDFIELDS LOCATION MAP

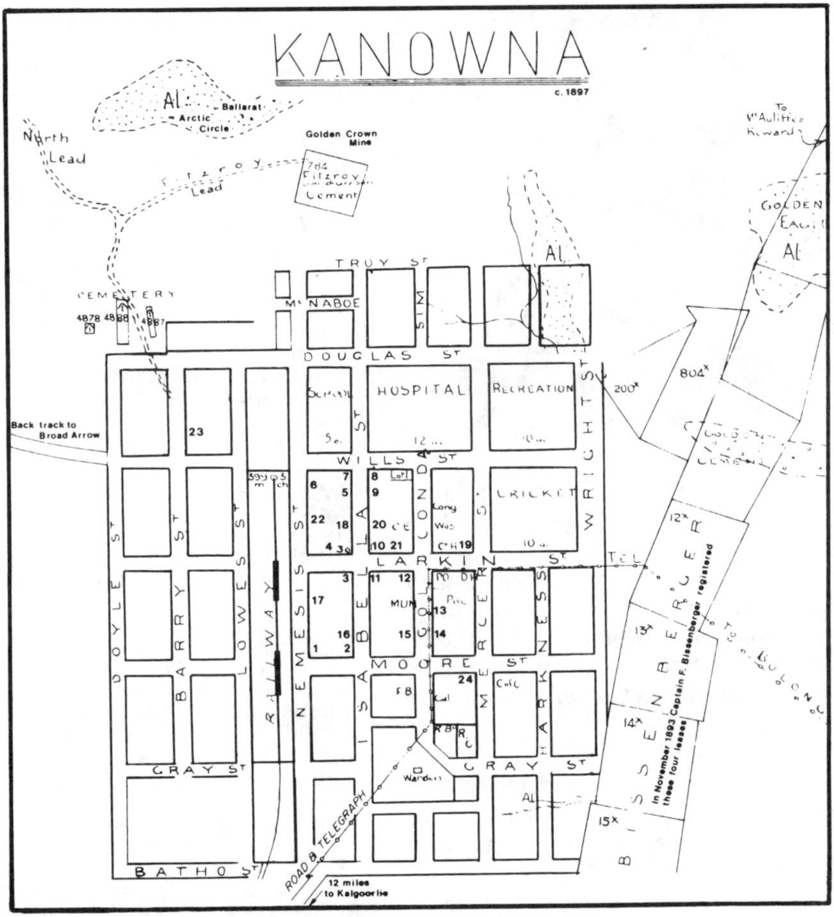

LEGEND TO KANOWNA'S SIXTEEN HOTELS; PROPRIETORS

1 Railway Bar (Mrs Violet Cook nee Thompson played piano here); Taylors.
2 Australia Hotel; Wilson Dunne, Holmes, last licensee Henderson, owner W.H. Trythall.
3 Criterion Hotel; Donnellan.
4 National Hotel (originally Walsh's Wine Saloon); Joe Solomon.
5 Australian Bar.
6 Sunbeam (Morans); Douglas Howard.
7 Occidental; Cutbush and Fimister/ Wilson's
8 Kanowna Hotel; Doyles.
9 Bathos Club.
10 Royal Exchange Hotel; Johnson.
11 White Feather Hotel; Coles/ McManus/ O'Donnells.
12 Federal Hotel; Wyatts.
13 Victoria Hotel; Mrs. Littleton.
14 Pickerings Hotel. Plus two hotels at Golden Valley three miles north. Plus two breweries McCabe's and Elliott's.

Other Addresses

15 Catchpole's Bakers Shop.
16 Tracey's Restaurant.
17 Baugh's Chemist.
18 Union Bank.
19 J.U. Smith's.
20 W.A. Bank.
21 Lowe's.
22 W.G. Manners' office and workshop.
23 W.G. Manners' first Kanowna home.
24 W.G. Manners' second Kanowna home (neighbours Father Long and Westcott and Timmel Families, house sold to W.G.M.'s brother Robert in 1900).

John Meiklejohn, the young Scottish prospector who discovered gold at Kanowna in 1893, and pegged the Golden Crown leases.

His remarkable trip to London with pockets full of nuggets, his meetings with company promoter Whittaker Wright and the subsequent prospectus float for the Golden Crown, is waiting to be written as another story of these exciting times.

The Golden Crown gold samples were displayed in a jeweller's window in Queen Victoria Street as part of the pre-float publicity.

This immediately brought the London traffic to a halt, causing the police to request the gold's removal from display.

Shares in the Golden Crown Company quickly ran up to £3.10.0 but the mine could not sustain it's success.

John Meiklejohn, also a veteran of the South African war, later went to the Klondike and became the proprietor and editor of the only daily paper on the field.

After Klondike he prospected for oil in Madagascar.

Prior to World War I Meiklejohn was a very wealthy man but the war destroyed his fortune and his later years, through to his death in February 1940 were spent in Perth where ill health brought him to a sad ending.

* * *

KANOWNA had 16 hotels and two breweries but a better way of describing the colourful past of the old mining town is in terms of its characters.

Names that will be bandied about when the ghost town's past is discussed during next weekend's 75th anniversary celebrations are certain to include . . .

John Meiklejohn: He found the Golden Crown mine, which provided wonderful specimens of gold. Whenever Mr Meiklejohn showed these to rich Englishmen in the halls of nobility back home a new mining company was formed.

Dr J. J. Holland: He performed the first major operation on the Goldfields.

W. G. Manners: He went to the Golden Crown as its mining engineer but discovered that no machinery existed. His grandson still runs W. G. Manners and Co. in Kalgoorlie.

Dean Moore: When the future Dean of Perth arrived in Kanowna from Ireland in 1898, the reception committee took him to a hotel and told him: "It's a bob in and the winner shouts."

A Mr Wyatt, who ran Wyatt's Hotel. He laid out a bowling green complete with ditches and mounds next to the hotel but no-one ever bowled on it.

F. X. Bissenberger: A Hungarian geologist nicknamed "busy-bugger" who was busy enough to find a reef running south that yielded 1,000oz. of gold in a few weeks.

Kalgoorlie Miner 1968

16

Kanowna (1895 - 1899)

On arrival at the Golden Crown I met Jonathan Bray, the manager. He was an old bachelor, and very fond of whisky. I soon found that he was more often drunk than sober, and quite a good fellow while he was drunk. But, when he was suffering a recovery, he was quite unbearable.

No machinery had arrived on the Mine, so that I had very little to do for some weeks but make myself at home.

One day Jonathan came to me and asked me to come up to his room. When I went there, he had a case of whiskey broached on the floor, an empty and a half empty bottle on the table. He asked me to have a drink. I joined him, then talked for a few minutes and we had another drink. But when it came to a third I refused, saying it was no good to me and that, if I took too much, I would have a headache next day.

"You could come in and have a pick-me-up." Jonathan said. When I informed him that having too much one day, I could not look at it the next, he said, "Well, boy, while you're like that, you're all right. But when you can get drunk one day and want a reviver next morning, you're settled. It's the same old drunk - the same old drunk."

So it was with Jonathan. That case lasted something over a fortnight, and it was the same old drunk all the time.

We had our meals at the Kanowna Hotel which was run by the famous Tom Doyle. It is supposed that Jonathan was worth at least £9 a week to Tom.

Tom was a broth of a boy from Ireland, popular with everyone. He was very rough, very vulgar, but very generous and open-handed. He is the hero of many Irish bulls and, if any real good bull is told, it is attributed to Tom Doyle.

Of course there were a lot of rough characters who used to come about, were not choice in their language, or particular as to whose company they used it in.

Tom had a barmaid named Kitty. As she was serving in the bar, two toughs came in from outback and, while drinking, used language of a very sultry and disgusting order. Kitty was shocked. She went out of the bar, found Tom Doyle, and complained that if the men were allowed to use such language, she could not stop in the bar.

Then Tom replied in great indignation, "Show me the... and I'll... soon put the... out."

Tom's language was exactly similar to that complained of, but Tom was unconscious of the humour of the situation.

Kanowna was quite a promising place at this time; there were quite a number of mines being equipped with machinery. I soon learned that the Golden Crown was endowed with a lode about 6" wide worth 10 dwts per ton and was being equipped with machinery purely for flotation purposes. There was quite a boom in Kanowna in those days[1], but it was never one quarter as lively as had been the Broken Hill boom.

I became friendly with a manager of an adjoining mine, his name was Harry

White Feather hotel - Kanowna (Publican D.J. "Dinny" McManus)
In mid distance: White Feather Livery Stables, "Horses & Buggies For Hire"

Presented by T. Doyle Esq. First Prize - Kanowna Handicap
Wheel Race - Two Miles Won by D.J. McManus
(Note engravers spelling error)

Snorthey. He was a Cornishman and supposed to be a miner. But he never went down the mine, leaving all the sampling to an underground manager. This was one of Kauffman's mines also.

Harry told me that he was doing very well; he sent all his salary home, and lived on the commission he obtained from the merchants.

His weekly reports indicated a lode three feet wide, valued at two and a half ounces to the ton, and he was erecting a ten head battery. When the battery was erected ready for running, Henry ordered a great spread and invited the people of Kanowna to celebrate the starting of the battery. He sent the bill, £30, to Head Office, but it was returned to him with the remark that the Company did not provide for these functions, he must pay the bill himself.

The battery got going with a great flourish of trumpets. It crushed for a month and cleaned up two and a half dwts to the ton. This was rather a drop from the estimated value of £10 to 10/-. Harry explained that the plates absorbed a large quantity, and that the rest was in circulation in the plant, but they would get it all next clean up. So he was given another month's run. At the end of that time, the result was the same, two and a half dwts per ton.

It appeared that the underground manager was one of those fools who estimated the value of the mine on the rich samples only, and disregarded the duffers. Because this was a mine with patches giving one good sample to a hundred duffers, his estimate was a bit optimistic. Harry was one of those fools who had not learned that if you want a thing done well, you must do it yourself. By such fools as these mining is muddled, capital is wasted and, often, false booms created, whereby the public are fooled into speculating and losing their money.

I only stayed about twelve months with the Golden Crown then started in business as an engineer and contractor. This was not a good move as events turned out, which fact may be attributed more to bad luck than bad management.

All business men were doing well at this time, but they were all putting their hard earned money back into mining and losing it. After much experience, I believe that luck in mining is a greater factor in success than knowledge. Not, of course, in running a mine when it is found, but in finding the mine and selling or keeping it when found.

I flatter myself I know something about mining, but I never got an ounce of gold in all the brightest days of Kanowna, while others who had never seen a mine in their lives before, would drop right on to riches.

Tom Doyle's brother came out from Ireland. He was a farmer, who had never seen a mine in his life, perhaps had never seen native gold before. He pegged out a lease on Kanowna Hill and, in less than three months, sold it for £2,000, then went back to Ireland to buy a farm.

A Johnnie arrived on the 'fields, he was straight from England, dressed with a splendid suit and leggings. He talked with an Oxford accent. He carried a nice new fossicker's dish, a nice new pick and shovel, and he wore an eyeglass. He came to some prospectors who were working an alluvial patch in a gully and remarked, "Oi say, you fellahs, can you tell me whar I maight faind some gold?"

The appearance of the Johnnie, and his method of expression, left no option with the experienced prospectors but to take a rise out of him. No knowing gold seeker would look on the top of a hill for alluvial gold, so they told him to go right on top of yonder hill and search there.

MR. J. L. MARTIN, MAYOR OF KANOWNA.

LARKIN STREET, KANOWNA.

ISABELLA-STREET, KANOWNA.

MRS. J. L. MARTIN, MAYORESS OF KANOWNA.

THE LATE WILLIAM ALEXANDER BLACK, accidentally killed at the Ivanhoe Mine, October 21, 1913.

ISABELLA-STREET, KANOWNA.

ISABELLA-STREET, KANOWNA.

"Oh thaink you," and off he went. In less than half an hour he came running back smiling and said, "Oh thaink you vera much, maites, look what oive got." and he held up a sixty ounce nugget.

Of course the fool did not peg out a claim, but the prospectors soon had the whole hill pegged out and, much to their surprise, found quite a lot of gold.

After leaving the mine I built a house and brought my family over from the East. We had four children by this time. I sent my brother Charles to meet them at Albany then bring them on to Kalgoorlie. I met them at the station with a double seated buggy and pair and, Kalgoorlie being the terminus of the railway at this time, the traffic had cut up the roads, so that they were a foot deep in dust.

My wife had brought her sister with her to help with the children and, before arriving at the station, they were all washed and combed and dressed in their best to look their sweetest when they met their daddy.

We got into the buggy. Before travelling 100 yards, so much dust had been thrown up by the wheels that the shining faces and pretty frocks were changed to a dull brown, and the happy smiles of the ladies changed to tears. However, we had to make the best of it, there were twelve miles of similar roads to negotiate. On arrival it was almost a matter of digging them out.

They thought all their clothes were ruined but, on hanging them on the line for a day, such was the nature of the dust that it all shook off. Their clothes served for many a day after. This might be a process of dry cleaning, or in other words, dry blowing.

We Tothersiders who had been accustomed to plenty of water and used the tin dish, tub, cradle and sluice to save gold from the soil, had to learn new methods here. The chief of these was dryblowing for gold. This process in its simplest form was accomplished by means of two iron dishes. The dirt was broken up small with the shovel. The dirt being dry, this was easily done. One of the dishes was filled with the dirt then poured slowly from a height of about four feet into the other. In passing through the air, the breeze carried the lighter particles away, leaving the coarse sand and gravel, and any gold that may be present, to fall into the second dish. The second dish is then lifted, the empty one placed on the ground and the process repeated until only coarse particles and gold are left. Then, with a swish of the dish, the gravel is spread out. Any specks of gold can be easily seen and picked out.

An advance on this process is known as the shaker. It consists of two or more sieves superimposed, and a hopper mounted on a frame having four slim legs made of hickory or kurrajong wood. The sieves have rifles in them, and the legs, being light, the sieves are easily reciprocated horizontally. The broken dirt is put in the hopper and the shaker is reciprocated. The dirt passes from the hopper over the superimposed sieves. The coarse gravel is retained in a screen and thrown out. After each operation the gold and some sand is retained in the rifles and, at the end of a run, the gold is extracted by means of the two dishes as previously described.

The best machine was the Lorden dryblower. It was similar to a shaker but had a very fine canvas screen at its bottom. The fine sand was kept lively by means of air passed under the screen with a bellows. This was capable of saving very fine gold. If an alluvial deposit had been treated through a Lorden machine, there was very little gold left for the next fellow.

The question of mining for alluvial gold on mining leases was then agitating

Kanowna Cemetery Rush, when "God's Acre" was worked for Gold. - 1897

White Feather Main Reef, Kanowna - 1904.

the public mind[2]. The lease holders contended that miners had no right to work alluvial deposits on leases, while the diggers believed that the leases were granted only for quartz and lode mining. This was not clear in the Mining Act, and Mr Wittenoom, who was then Minister for Mines, brought in a bill limiting alluvial mining to ten feet deep. This measure caused a riot among the miners and brought Mr Wittenoon into great ridicule, earning him the sobriquet (nick-name) of "ten foot Ned". Even Sir John Forrest, the great Sir John, got a poke in the ribs with an umbrella during the excitement. The Act had to be repealed, so miners won the right to mine alluvial gold to any depth, wherever it was to be found.

Prior to the settlement of this question, two prospectors named Sim and Gresson were working a six acre lease in Kanowna, getting good gold from a surface deposit. We did not know at that time whether it was alluvial or lode. It was a cement conglomerate and very rich.

The Ballarat lease was adjoining on the North. I suggested to a friend that we peg out an alluvial claim on the end of Sim and Gresson's lease and see what the Ballarat leaseholders would do. When we got there we found that Morris and party had pegged out six men's ground. Instead of pegging on to the end of Morris and party, we went away, disgusted, remarking that we were too late.

I had work to do in Widgiemooltha at this time and went away for a fortnight. During my absence Sim and Gresson got a 100 oz. nugget and Morris and party a 20 oz. nugget. This created a rush. My friend pegged out on my behalf, but we were half a mile away from where we should have been. On my return I noticed a shaft thirty feet deep that had been sunk by the Ballarat Company. The dump was a white kaolin with sand interspersed. It was not taken up, and adjoined our ground on the west. I sampled the dump carefully but did not get a colour of gold, so decided to keep on where my friend had pegged.

A week or so later a party took up this shaft and, after working a few days, struck rich gold. It proved the richest claim on the lease and was called the Arctic Circle.

We never saw the colour of gold in our claim. Later on I bought a claim further north. The claim on my south boundary was yielding good gold. After working this claim for some months it was discovered that the lead turned at right angles and went away west, just outside of my boundary.

This was the third time my luck turned me down. So I never beheld a colour of gold in all that great and wonderfully rich deposit.

It was while this lead was being worked that Father Long saw that wonderful sickle nugget. This was the sensation of the year. The lead had brought many thousands of miners to Kanowna. But, when one morning Father Long announced that on the previous night he had seen a wonderful nugget of gold worth £1200, it was in the form of a sickle and had an ironstone handle. Excitement became intense. He alleged that the two men had brought it in and shown it to him, but had bound him to secrecy as to their identity or the location of the find.

When this news was published in all the papers, thousands more rushed to Kanowna from all parts of the fields. After a day or two of intense excitement they began to clamour and press Father Long for more information. The manager of the *Democrat,* the Kanowna newspaper, run by a Mr Cohen, approached Father Long and was informed that he, Father, had seen it again, that it was in the Commercial Bank at Kalgoorlie.

The crowd gathered outside Donnellan's Criterion Hotel to hear Father Long indicate the locality where the Sacred Nugget was found. 11 August 1898.

Father Long at Kanowna - 1898

Mr Cohen went to Kalgoorlie, but the Manager of the Commercial Bank knew nothing about it. He then inquired at all the other Banks, but they were equally ignorant. This cast some doubt on the matter. But Father Long was a priest so no one could doubt his word.

The men were getting out of hand and, after several days of anxiety, a mob invaded Father Long's camp and threatened to lynch him on the spot unless he told them more about the nugget.

It is difficult to say what would have happened to the Father had not Warden Troy, a highly respected and strong man, intervened. He quietened the crowd and got them to leave, promising that he would use his influence with the Father to disclose the secret.

The result was that Father Long agreed to disclose the location of the find, with or without the owners' permission. He promised to announce it from the Balcony of Donnellan's Hotel at 2 o'clock the next afternoon.

There were thousands of men there at the appointed time, all eager and full of excitement. Then Father Long declared that the nugget had been found three miles from Kanowna, on the Kurnalpi Road, just this side of Lake Gwen. No sooner was the announcement made then a stampede started: they flocked out in carts and buggies, on horseback, on bicycles, on foot, crowding the road and jostling each other in their eagerness to get past. Before the day was out thousands of claims had been pegged and, next day, thousands of men were working with shakers and tin dishes. But never a colour of gold did they see. Father Long was soon after removed from Kanowna and, although he was then a fine, young, healthy man, he was dead within a few months. Poor Father Long, he had really seen the gold in the shape of the sickle, but had got a wrong interpretation of it. I can remember clearly the occasion.

The Arctic Circle was having its fortnightly clean up. It was to be a record one. Tom Doyle was a shareholder and, in anticipation of the big dividend, opened a bottle or two of champagne.

Father Long was on the scene at the time and joined with us in the festivities. Bottle after bottle was opened by the lucky shareholders and, when the gold was ultimately smelted, was brought to the pub in five bars of gold arranged around a rusty iron prospector's dish. They were thus in the form of a sickle, with the iron dish to approximate the ironstone handle. Its value was £1200.

When Father Long awoke next morning, he had forgotten all about the celebrations at Tom Doyle's but had a hazy idea that he had seen the nugget, or perhaps he dreamed it, and, having once made the announcement, felt that he had to stick to it, to see it through to the last. Thus had booze claimed another victim and destroyed another life.

Of all this gold that was obtained from this wonderful lead, the greatest proportion was wasted in drink. Many of those who had made fortunes at that time were cadging for a drink a few years later.

I knew one party who took cases of lager down their mine and, after working an hour or so, would spend the rest of the morning or afternoon drinking. They took out a lot of gold but, when their claim was worked out, they had scarcely a penny left.

A lot of money went into backing prospectors. Tom Doyle backed anyone who put a proposition to him, with the result that, although he made one

The Kanowna Express arriving at Kalgoorlie - 1898

The Kanowna Railway Station - 1904

fortune on the lead and another in his pub, he was ultimately stranded. He committed suicide, without a penny to bless himself. A lot of these alleged prospectors were loafers who were living on the £2 a week provided by their backers, doing as little work as possible.

I had been appointed manager of the Kanowna Carbine Mine. It was worked out so we were public crushing. Mr Con Long, a member of the merchant firm of Long and Seebeck, came to me and arranged for a fifty ton crushing. He told me he expected an ounce to the ton.

He was backing two men who showed him very nice prospects of gold every time he visited the mine.

About a fortnight later I saw him again and asked him when his crushing was coming along. He said the men wanted to make it up to sixty tons. About a month later I met Mr Con Long again and asked him about the crushing. He informed me that when he insisted on the crushing being sent, they promised to do so. But, when he sent the drays out, the men had disappeared. And, on sampling the dump, he found there was no gold in it. They had sponged him for six months. So it was with many others all over the fields. All sorts of schemes were practised to beat the other fellow. One had to know something about the business to avoid being taken down.

It had been the custom to scrape the copper plates only, thus a little of the prospectors gold was always left on the plates as a nest egg for the battery owner at a later date.

One battery manager had introduced the system of rubbing the plates with a flat piece of soft wood and sharp sand, by means of which a great deal more of the gold was obtained. I had not, at this time, adopted the practice and, in consequence, a little amalgam had accumulated on the plates since I previously cleaned them.

Some prospectors who had been crushing with the other battery and taken all their good crushings there, but had fifty tons of what he called headings, that is, the gravel immediately above the true wash. This was poor, averaging about seven dwts per ton.

They came to me to make arrangements to put this through for them, stipulating that they be allowed to sand the plates. This I granted, and instructed them to bring their crushings along the following week.

They examined the plates and were very pleased with their appearance. I guessed their intention, they wished to make their poor crushing into a good one by the gold off my plates.

So, in the meantime, I took the plates up and treated them with washing soda. This process is very little known. The plates are washed over with a saturated solution of washing soda then heated gently until the soda "sizzles", being careful not to evaporate the mercury. This softens the amalgam so it can be scraped off with a table knife. I took off thirty two ounces of gold.

When they returned they examined the plates again. They looked just as good as ever but, when they cleaned up, they had just eighteen oz. of gold.

They were disappointed; they fully expected to get twenty five or thirty oz. off my plates.

During the Alluvial Lead Rush, Kanowna was a prosperous town, it had eighteen public houses, all making fortunes. It is a failing with most people when they get a heap of money for the first time, they think it is inexhaustible, and just throw it away.

I have seen three or four men come into town after selling a mine or having a rich clean up, go to a pub, get a private room and drink champagne by the case. The landlord would collect all the girls in the house and put them in the room with the fools. The girls would drink very little, but take sly opportunities to pour it out on the floor then have their glasses filled again. When the prospectors got drunk the champagne bottles would be filled with challis and soda and sold at 25/- per bottle. At one of these sprees I have seen a dozen bottles taken into a room within an hour.

One Sunday a Wesleyan parson expressed in his sermon his indignation at the fact that there were seventeen houses of ill fame in the town. These were occupied by all nationalities of women, French, German, Japanese, Russian and English. A mother and her young son, who was a telegraph boy, was returning home from church when the mother was astonished at the boy saying, "Mother, the parson was wrong tonight."

"In what way was he wrong, Johnnie?"

"He said there were seventeen houses of ill fame in Kanowna, but there are only thirteen, mother."

The mother was shocked to think her boy knew so much about houses of ill fame but, being discreet, she thought it best to find out how much he knew, so she asked, "How do you make that out, Johnnie?"

"Well, mother, you count: there are Doyles, Donnellans, Bathos, Cutbushes, Johnsons," and so on, enumerating all the saloon keepers. His mother was greatly relieved, but the boy was maybe not far wrong after all.

When the population had increased to a certain number Kanowna was proclaimed a Municipality, so nominations were required for six councillors.

There were twelve nominations, and there was an election. Tom Doyle was one of these and, at the declaration of the poll, all the candidates spoke a few words. Tom tried to get out of it, but was ultimately induced to mount the beer case rostrum and, with his simple grin, started.

"Gintlemin, when oi' git in the council, the furst thing oi' will do, will be to shift all the sanitary arrangements two miles out of the town."

Cheers for Tom, and much laughter.

"And, Gintlemin, I have always done my best for the town while I have been in it and, whin I am in the Council, I'll do as much in the past as I've done in the future."

More cheers and more laughter.

Then Tom added, "Come up ye and have a drink."

More cheer, followed by a general adjournment to the pub, where Tom shouted everyone. To appreciate the full humour of this speech, one needs to see Tom's simple grin and hear his rich brogue.

Later on we had a Mayor named Tonigees, a Dutchman, who was studying English from a dictionary. He was quite a Mr Malaprop. One night, when publicly reading a letter, he said, "Genelmen, I have been depicted to retaliate this letter to you." then read the letter.

Another time he was proposing the health of the Government, whose Ministers were on a visit of inspection to the district. He said, "The Ministers have seen all our mines, they have acquisitioned some specimens of our ores, which they will be able to place in their aquarium."

So these mushroom towns discover genius and make heroes of men who would otherwise never have been heard of.

1. Appendix IV gives details of the 1894 float of White Feather Proprietary Gold Company Ltd.

2. Appendix V summary pages from the 1897 Royal Commission to resolve alluvial working on mining leases.

MINERS' INSTITUTE.

SUNDAY NIGHT.
SUNDAY NIGHT.
at 8.15 p.m.

GRAND CONCERT.

Orchestral, Instrumental, Vocal, Choral.

The following Talented Artistes will appear :—

HARRY FITZMAURICE.

Mr. Dan Cusack	Mr. Harry Gilbert
The Destree Bros.	Mr. G. Keenan
Mr. Lynch	Mr. J. Sheehan
Miss Florence Hope	Miss K. Moran
Miss M. Moran	Miss Ettie Hamilton

and

The KANOWNA ORCHESTRA
of Twenty Performers.

Musical Director : Mr. Geo. Gardner, R.A.M.

Popular Prices—2s., 1s.

Kanowna. August 21, 1898

The arrival of miners at Kanowna - 1897

The Kanowna Post Office - 1898

Kanowna's "Barrowman" James Balzano, arrived in early 1895 and lived in Kanowna, recording its history till his death in 1948, aged 89.

James Balzano in action on his Kanowna/White Feather lease. Photo taken by M. Crofton & Co. of White Feather at 10.52am May 11, 1896.
Balzano's remarkable life is detailed in two other books:-
"Kalgoorlie - The Golden Years to 1966" By Charles & Nancy Manners.
"Kanowna's Barrowman - James Balzano" By George Compton & Ron Manners.

COOLGARDIE
Laverton
Paddington
Lawlers

Cumberland St. Kookynie.
Menzies.
Kookynie.
Malcolm.

WHAT LAY BENEATH THE SANITARY DEPOT?

Little did the old timers know that 90 years after the "decline of Kanowna", a spectacular gold discovery would put Kanowna back on the map.

The 1990-91 discovery of Kanowna Belle by joint venture partners Delta Gold N.L. & Peko Gold Ltd has shown how modern prospecting methods can penetrate the 40 metres cover of barren overburden, thus revealing the treasure undetected by earlier explorers.

The other reason for Kanowna Belle remaining hidden is aptly described by journalist/mining historian Ross Louthean "This area was obviously anathema to the early Kanowna miners, because the old night-soil cart track goes straight over the orebody."

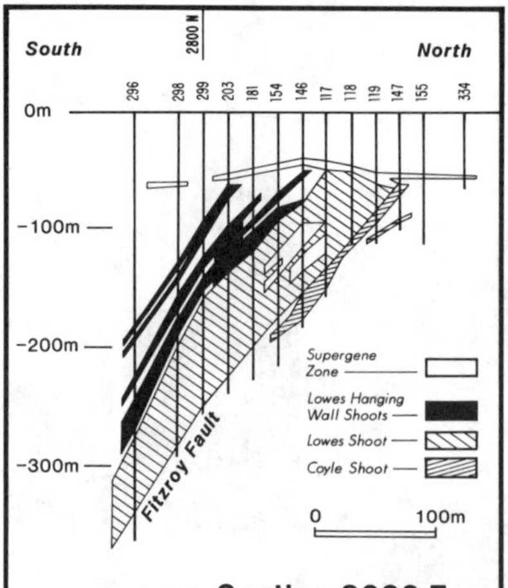

Kanowna Belle **Section 8600 E**

"Based on drilling completed by September 1991 The interim total measured and indicated resource at Kanowna Belle is now estimated to be 11·2 million tonnes at 5·8 g/t gold (uncut), containing 2million ounces of gold"

Delta Gold NL release to ASX 25-9-91

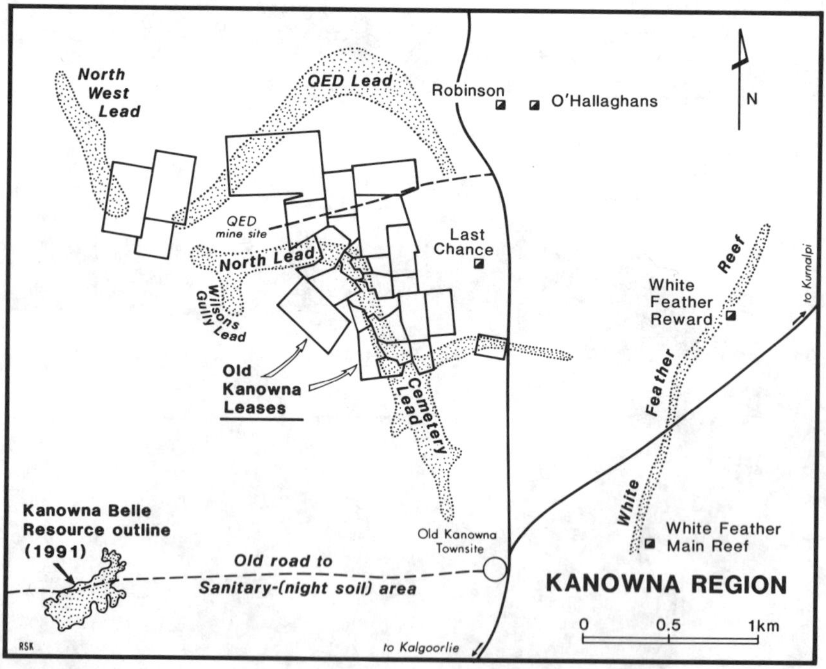

17

Kanowna's Decline (1900 —)
(The Move To Kalgoorlie)

As the lead became worked out and no new discoveries were made in the Kanowna district, the town began to wane.

All those who had not been fortunate enough to strike a rich claim, and all those who had, but had spent the money on booze and were hanging to the place hoping that any day would reveal a new strike and give a new lease of life to the town, were becoming poorer and poorer; gradually one business man after another was forced to give up the struggle to seek fresh fields. The pubs hung on the longest. The monopoly value of a licence induces this state of affairs, and we had the anomaly of about ten or twelve hotels. But, in all other business, things were gradually reduced until there was only one of each. At last the pubs had to go because the population disappeared. Today,[1924] there is but one hotel, The Australia Hotel, the fortunate possessor of which is making lots of money, although the local population is less than 100. [This has since been demolished] The Municipal glory is gone and even the reduced status of a Roads Board has been abolished. The district has been included in the Kalgoorlie Roads Board area.

So it was with me and, practically giving my property away, in the very early stages of the decline, I was compelled to pack up my traps and move to Kalgoorlie, where being offered a job with the London and Hamburg Gold Recovery Company, became connected with the great Golden Mile.

The London and Hamburg G.R. Coy., was a German firm managed by Dr. Diehl and Mr Gunther. the latter being the engineer. They re-designed the Brown Hill and Lake View Consols plants while I was with them. (A firm by the name of Turrell[1] and Kay had the contract for most of their work). They were undoubtedly clever mining engineers who did very excellent work and introduced a number of new and useful ideas into the construction of machinery, such as flush headed set screws on shaft collars, protected shaft couplings, fixed bushes in loose pulleys to save wearing the shaft, chain oiling bearings, and other improvements on Australian practice. Although they both spoke English very well, with a German accent, they were very amusing at times in trying to express themselves.

Mr Gunther was trying to tell us that the best was the cheapest, but got mixed and said, "The dearest is the cheapest," when we laughed he tried to correct himself, "What you say den? The dearest the best, eh?" Then we told him what he meant to say.

Our English synonyms puzzled them very often. In conversation with Dr. Diehl, talking about his trip out from Germany, I asked him if he were a good sailor. "A sailor?" he replied, "But I am not a sailor, I am a metallurgist."

They had a great abhorrence of flies and had all the doors and windows in the offices covered with fly screens. However, in spite of their precaution, a fly had invaded their sanctum. They both occupied the same office, and I was greatly amused on entering one morning to see both of them with little rubber squirts filled with Insectabane, dancing and jumping all over the office after the offending fly and, as chance offered, squirting little puffs of the powder at it. I suggested that it would be better to catch the fly and one could hold it while the other squirted the poison at it, but they did not see the humour of my suggestion.

I was with this firm about twelve months when I left and, because Kay was going out of the firm of Turrell and Kay, I joined this firm. It became Turrell and Manners. We were very successful. Dr. Diehl and Mr Gunther were very much annoyed when I got the designing of the Golden Horseshoe 100 head battery, and an additional twenty five heads on the Lake View Consols, besides other big work. They would not speak to me at all. So started the first war between Germany and Australia.

Some very interesting and useful changes took place in mining and metallurgical practice during the early stages of development on the Kalgoorlie Goldfields. for Kalgoorlie was fast outstripping Coolgardie in mining importance.

The first managers of these mines were all Tothersiders whose experience, particularly at Ballarat, was with clean quartz, and practically pure gold. With the ordinary stamp battery and copper plates, and in some cases blanket strakes to save concentrates and black sand, an extraction of 90%, leaving less than 1/- worth of gold in the residues, additional treatment such as cyanide was unnecessary for, when the gold was recovered and smelted it was practically pure.

These men for the first time in their lives ran up against something they did not understand. The gold was found in all sorts of matrices and the extraction by amalgamation was only from 40% to 60%. In many cases as much as an ounce to the ton was being left in the tailings.

As far as the sands were concerned this presented no great difficulties because young metallurgists were available who understood the cyanide process. So vats were erected and the gold leached out of the sands but, in the soft ores, a great quantity of slimes was formed which could not be leached and was, therefore, not amenable to this treatment. These slimes represented nearly fifty per cent of the residues, with the result that large quantities of valuable slimes were being accumulated in the dams on the mine.

Mr J.W. Sutherland, then a young man, who had been educated at the Ballarat School of Mines, and had several years of Broken Hill experience, conceived the idea of treating these slimes by means of a filter press. He was then Metallurgist on the Lake View Consols Mine, controlled by Mr Chas Kauffman. He put his ideas before Mr Kauffman and was authorised to install a small experimental plant. His experiments were so successful that he reported favourably and was asked how much money he would require to install a commercial plant. when he replied "about £5,000," Mr Kauffman said "Well, boy, there is £10,000 placed at your disposal, let us see what you can do."

Thus, the first filter press plant was installed. It proved so successful that it revolutionised the treatment of ores altogether. Many of the mines installed tube mills to slime the whole of their ore so as to get the better extraction obtained by this process.

The London & Hamburg Company introduced some small improvements in this process and, I believe, tried to claim the credit for the invention. But I know personally that the credit for this great discovery in slime treatment was entirely and solely due to the genius of this young Ballarat Metallurgist, J.W. Sutherland.

* * * * * * * * * * *

I remember when I was quite a young man, my father saying that if one could not see gold in the stone, or get a prospect in the dish, there was no gold in the stone. I believe this was quite true as far as mining practice went in Ballarat.

Many of these old Managers evidently brought these ideas to Western Australia and, being old and set in their ideas, were hard to convince otherwise, with the result that assaying was not very extensively adopted. In consequence, many lodes and reefs were missed. Some lodes in which no free gold was visible, and which gave very poor prospects in the dish, were proved by assay to contain quite payable quantities.

And so it was when the sulphide zone was reached: the quartz-streaked diorite with splashes of what looked like mundic was thrown in the mullock dump. Many hundreds of tons were thus thrown away. Then a scientist, who had experience in other parts came along and recognised the supposed mundic as Telluride of Gold. It was discovered that we were the possessors of some of the richest mines in the world.

It was Mr Holroyd who discovered the presence of Telluride in the Kalgoorlie lodes in 1896.

The old-school men had to pass out and make room for young men with metallurgical knowledge, for here we had a chemical problem which had to be solved and it required men who were not too old or too wise to learn new ideas.

Two methods were soon introduced, one being a dry mill process in which the whole of the ore was roasted, the other being by wet mills and concentration. The ore being treated by the bromo-cyanide process and the concentrates roasted. On these principles some very fine plants were erected; at that time Kalgoorlie contained some of the finest and most up to date plants in the world.

The mines on the Golden Mile proved very rich. Big fields and big dividends were the order of the day. The Lake View Consols Mine, when they got their plant erected, were producing as much as a ton of gold per month, and several others were returning as much as 20,000 ozs per month.

Unfortunately for Australia, the Adelaide mining people who originally owned these mines sold them whenever a tempting offer was made by British investors, who were more prepared to take the risks, so most of these wonderful dividends were sent to London to enrich that home of wealth and imporverish Australia. Australia got the wages, which were none too high, and the paltry rents charged for these mines of fabulous wealth. Good salaries were paid to the managers of these big mines, and to some of the staff officers who held very responsible positions. Large as these salaries were, they were not in any way excessive for the few good men who were endowed with a large share of ability and commonsense.

Of course mistakes were made, much machinery was erected and afterwards scrapped. These mistakes were due to lack of experience. This experience had to be bought so such mistakes were quite excusable. But there were some mistakes due to what might be called "swelled head," and were very costly. In fact, in one case, it would have paid the company to give several years' salary for the manager to stay off the mine altogether. The treatment of the ore was by means of roasting. It was under the supervision of the metallurgist who was without exception the cleverist and most up to date chemist in Kalgoorlie. The plant consisted of nine duplex Edwards roasting furnaces and, of course, other necessary plant. They were treating 18,000 tons per month, the residues were going out valued at twelve grains per ton or about 2/-.

The manager, in order to boost his reputation, wanted to increase the tonnage to 22,000 tons per month. He approached the metallurgist with a view to having it done.

The metallurgist refused on the grounds that they were now doing their maximum tonnage consistent with good work. The manager insisted on his wishes, and the metallurgist persisted in his.

Then, when the metallurgist had gone home for the evening, the manager went down to the plant and increased the feed to the furnaces. In the morning the metallurgist found a foul roast and asked the reason for it. On being told, he altered the feeders back to their original marks then growled at the manager.

The next night and the night following, the same thing occurred. The manager and the chemist nearly came to blows. The result however was that the cyanide solutions became so foul from the bad roasting that the residues rose to five dwts or 20/-, and all of the solutions, worth thousands of pounds, had to be thrown away. Thus was a loss of at least £36,000 worth of gold thrown in the dump, never to be recovered. And the loss in solutions and stoppage in plant possibly another £10,000, caused by the stupid ambition of a man who should have had more sense. Thus are many companies ruined, for they cannot all stand a loss of this sort and survive. But the fool managers still thrive and prosper and, let us hope, learn some common sense in time.

1. S.G. Turrell was a Department of Mines Surveyor in Ballarat in 1904, later operating in Kalgoorlie as an Assayer, Metallurgist and Analytical Chemist.

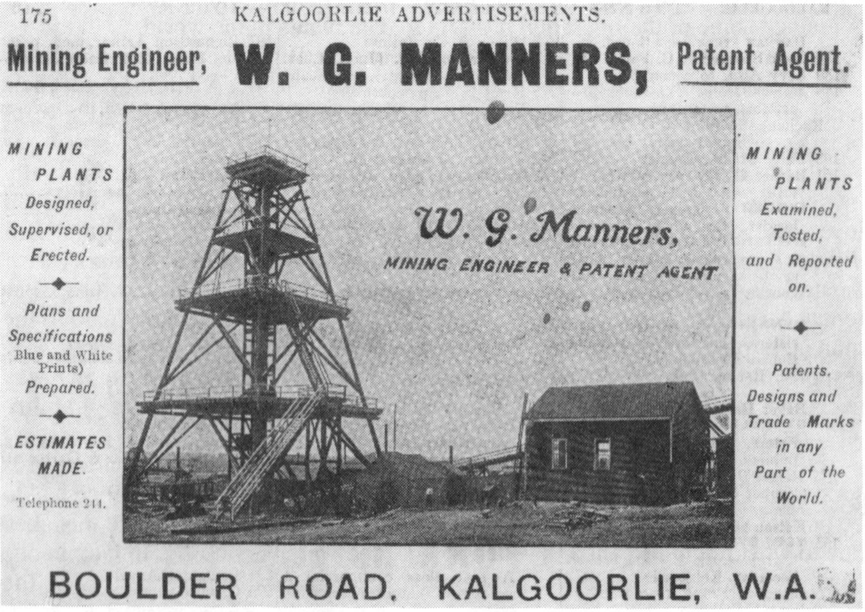

North Kalgurli - Original Main shaft c.1902

W.G. MANNERS
BRIEF SUMMARY OF MINING PLANTS DESIGNED AND ERECTED 1896 - 1924
(FURTHER PARTICULARS CONTAINED IN APPENDIX VI)

Golden Crown Mine - Kanowna 1895-6.
Golden Horshoe (100 Head battery, treatment and roasting plant) - Kalgoorlie 1901.
Brownhill (Redesigned plant) - Kalgoorlie 1902.
North Kalgurli Headframe (Original main shaft) - Kalgoorlie 1902.
Lake View Consols (Redesigned plant) - Kalgoorlie 1902.
King Battery (Complete plant) - Hampton Plains 1903.
North End Mine (Redesign expansion and install powerhouse) - Kalgoorlie 1906.
Sons of Gwalia (New 10 head battery, plant, boilers, assay office etc.) - Leonora 1907.
St. George Mine & Treatment Plant - Mt. Magnet 1908.
Cobar Mines (New plant) - N.S.W. 1909.
Cam & Motor Mines (New Mining and Treatment Plant) - Rhodesia 1911-12.
Ida H Mine & Treatment Plant - Laverton 1914-15.
Sundry redesign and relocation of existing mines and plants 1915-1924.

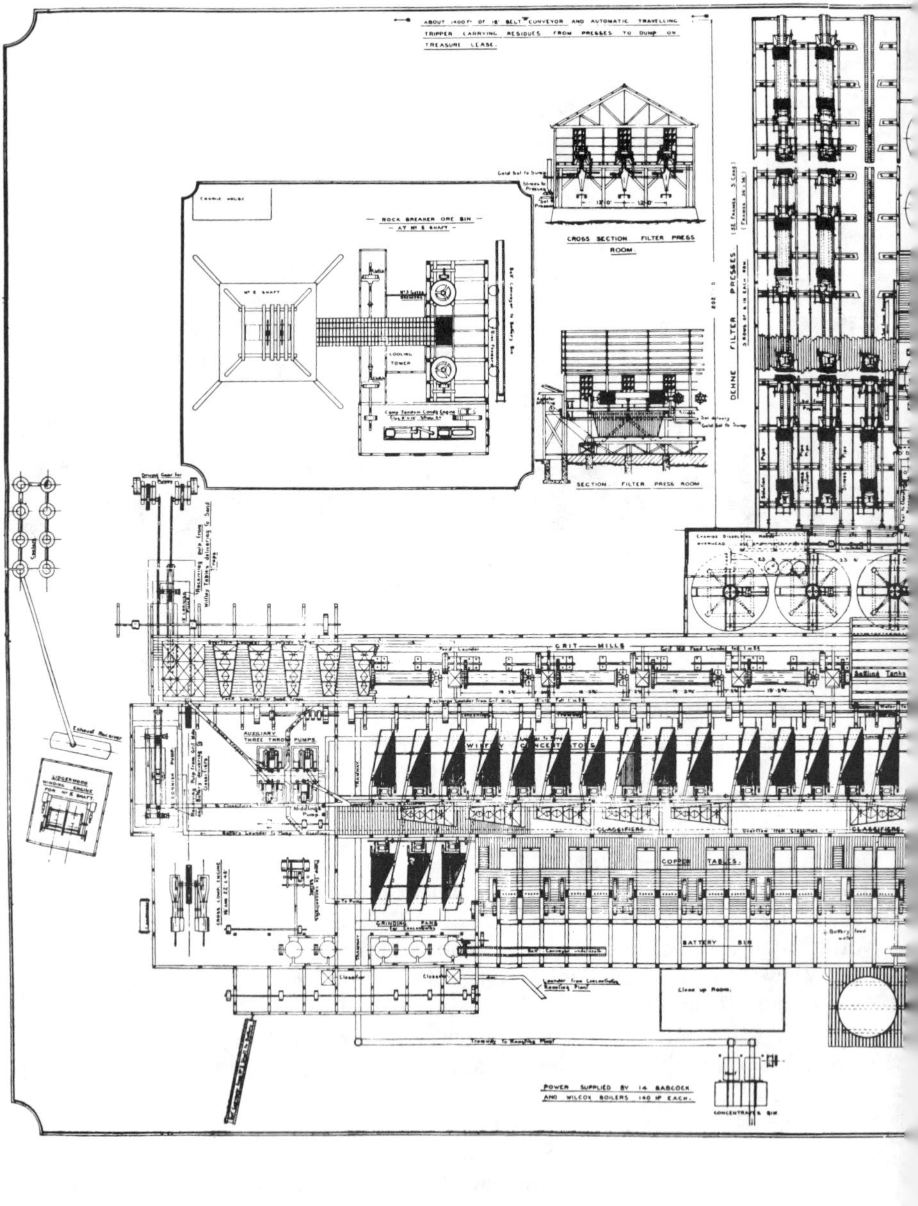

Golden Horshoe Gold Mine - Kalgoorlie 1901

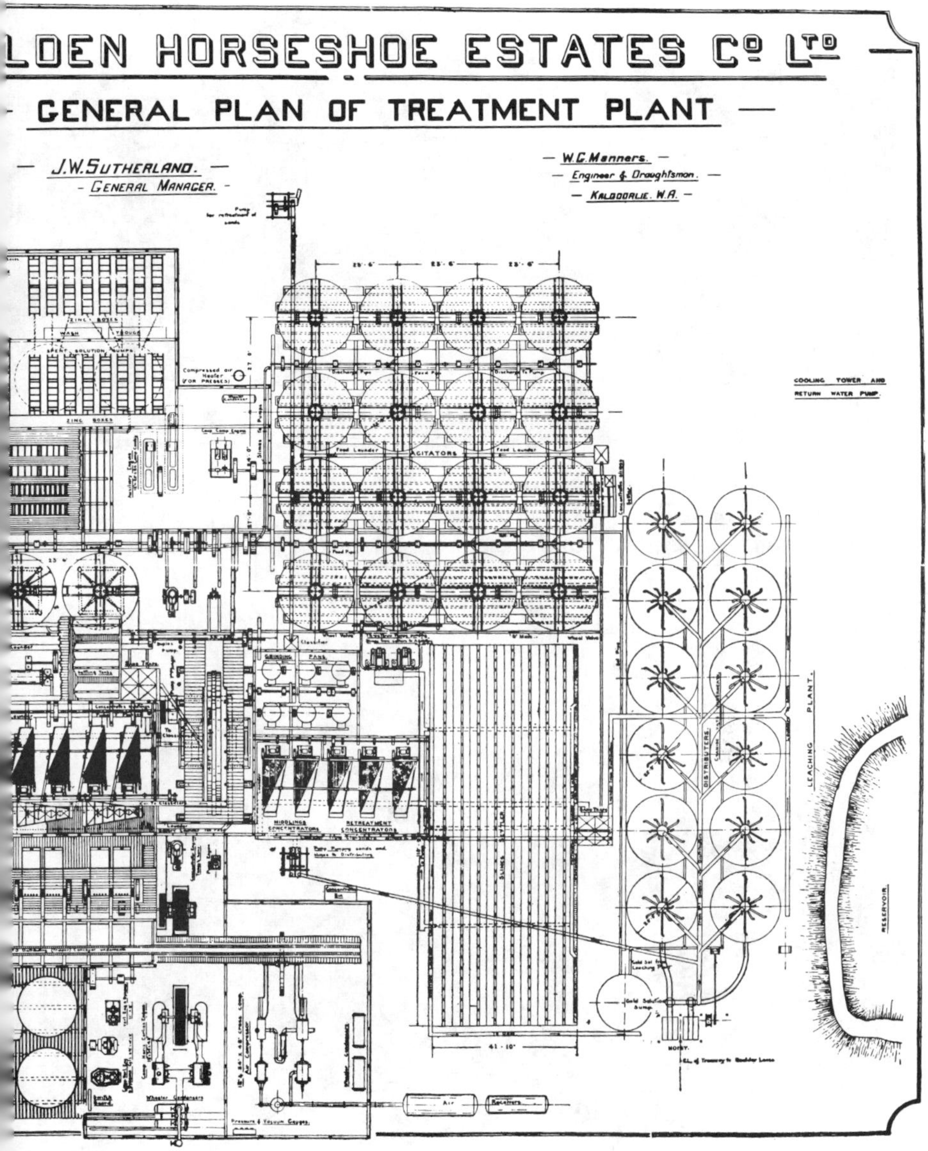

General Manager - J.W. Sutherland. Design & Construction - W.G. Manners.

Golden Horseshoe - from south west.
This gives a panorama of the lease from Main Shaft (left) to No.2 Shaft (centre), with treatment plant and roasters as designed by W.G. Manners.

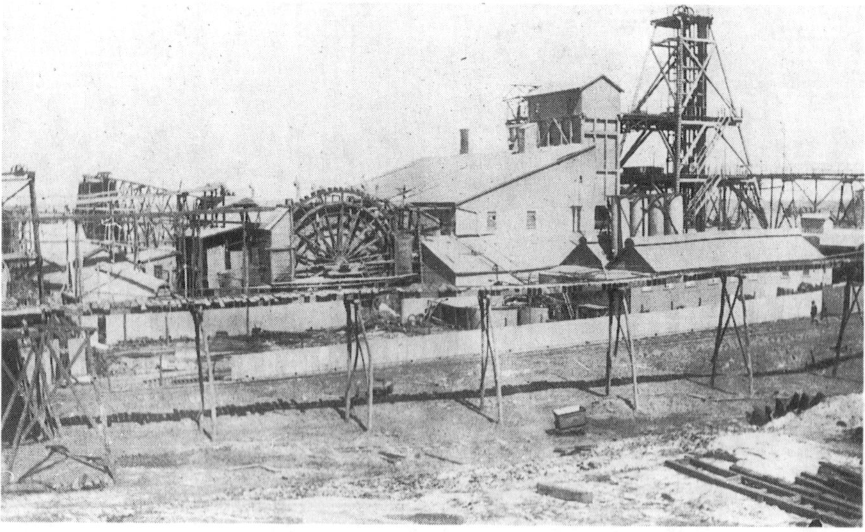

Golden Horseshoe - Portion of plant from south east. Note tailings wheel in No.1 Mill. C.1901

It also shows the Great Boulder Bleichert elevator and Lidgerwood cable-way. In the foreground is the Horseshoe tailings belt conveyor, elevated over the loop-line and main road. - 1911.

Golden Horseshoe - from the east. The Horseshoe Main Shaft is inside the fenced area, centre right, and the tailings wheel foundation from No.1 Mill is in mid-distance left. The costeans have been used to test a possible open cut area. Photo taken 1979.
(Editors note - 1991. This area no longer exists, having been mined as part of The "Super Pit").

CHAPTER V

GOLDEN HORSE SHOE MINES

ATTEMPTED WRECKING OF THE GOLDEN HORSESHOE ESTATE MINE

THE following chapter speaks for itself. It is given here as Sir John Purcell delivered the speech to the shareholders of the company. Nothing has been changed, except to leave out the statistics of the yearly report which would not be interesting to the reader. The company paid for its publication in several London newspapers, including the Financial Times, this copy is from the Statist, April 21, 1905. In this speech Sir John Purcell shows how the Hoover-Moreing firm tried ... of this rich mine, and when the lawful officers atter... ...shortage, Hoover told Sir John, that Joh... pro... mi... th... ... Of course, Sir

Purcell and Sir West Ridgeway, looking after the interest of their shareholders, we leave it to the reader to decide what this firm did to mines, in which they had sole control, without anybody watching them.

THE GOLDEN HORSESHOE ESTATES

"In March, Messrs. Bewick, Moreing & Co., were appointed consulting engineers and general managers in Western Australia. In the following September, the directors removed them from these offices, and Mr. J. W. Sutherland was replaced in full charge of the mine. The two points specifically referred to by the auditors in their report relates solely to the period during which Messrs. Bewick, Moreing & Co., were in charge of the mine. The bullion in question has since been surrendered. The auditors report states that:—

"All our requirements as auditors have not been complied with, and we report to the shareholders that we examined the accounts dated the 31st of December, 1904, *bearing our certificate and have required certain original invoices for stores purchased in Australia, which have not been produced to us, and that in the balance sheets there is included under the item 'Bullion and Sulphide ore in transit' the sum of* £14,349, 5s. 3d., *being the estimated value of the bullion deposited in other names than the company with the Union Bank of Australia at Kalgoorlie, for which no certificate has been presented to us."*

—London Times, April 19th, 1905, page 13.

The above is the report of the auditors of the Golden Horseshoe Estates, upon the item mentioned by Sir John Purcell, as to the bullion taken out of the mine by the Hoover-Moreing firm and placed in their own bank.

Pages 99 & 100 from James J. O'Brien's book "Hoover's Millions and How He Made Them" - 1932.

Cobar Mines - N.S.W. 1909

The Cobar Leader

PEACE PROGRESS, PATRIOTISM

WEDNESDAY. FEBRUARY 24, 1909.

Mr W G Manners, the consulting engineer and patent attorney, who lately come to Cobar in connection with the proposed new plant on the Cobar Gold Mines, has finished his work and proposes returning to Sydney about the middle of March. Inventors should take this opportunity of consulting him on patent matters, thus saving the cost and inconvenience of visiting Sydney. His long experience as a registered patent attorney, should enable him to safeguard the interest of investors, while a personal interview is often the means of avoiding errors. He will supply all information free of charge. Address, Cobar Gold Mines.

Cobar Herald
AND
NYMAGEE ADVERTISER.

"NON SIBI, SED PATRIÆ."

FRIDAY, FEBRUARY 26, 1909.

Inventors have an opportunity at the present time of obtaining full information in relation to patent matters, as Mr. W. G. Manners, consulting engineer and patent attorney who has just completed his business in connection with the Cobar Gold Mines, is remaining till about the 7th of March and will be pleased to supply intending patentees with all particulars required for obtaining patents in any part of the world. Mr. Manners has had a long and valuable experience as a mining engineer and patent attorney, and his presence in the town will be the means of saving trouble and cost to patent clients, who will find a personal interview a great advantage in safeguarding their interests. Present address, Cobar Gold Mines.—0

Kalgoorlie Miner.

TUESDAY, APRIL 27, 1909.

NEWS AND NOTES.

Mr. W. G. Manners, who returned from the eastern States last Saturday, has designed a new plant for the Cobar goldmines, and a reconstruction of the Occidental Goldmining Company's plant. Both of these schemes are now under consideration by the directors of the respective companies, and will be proceeded with early next year. Cobar is very quiet at the present time owing to the decline in the price of copper, the Great Cobar mine, and the Queen Bee, which are copper mines, feeling the decline most. The former, however, under the expert management of the lately imported American engineer, Mr. Bellinger, promises good results when the alterations to the plant are effected, and the process of treatment now being introduced is in full operation. The most interesting feature, so far as the field is concerned, is the Moore filtering plant in work on the Occidental goldmine. Several slight alterations introduced by the metallurgist, Mr. Chapple, have made this plant a great success, so much so that the small plant designed to treat 50 tons per day easily copes with 150 tons, and, in addition to treating the whole of the slimes produced daily by the battery, also treats 70,000 tons of accumulated slimes of an average value of 2½ dwt., at a profit of 5/ per ton. Owing to the oxidised ores being almost exhausted, and a slight increase of the copper contents in the lower levels, it is proposed to install tube mills and slime the whole of the ore.

Excavations for King Battery - 1901
Located midway between Kalgoorlie and Kambalda on Locn.48.

The completed King Battery, with a capacity to treat 100 tons per day. - 1902
The 20 head battery was fed by rail links to surrounding mines 8km SW (Merry Hampton) and 5km north (Hampton Boulder/Jubilee)

Erecting the building over the tailings wheel - 1901

ilings wheel as it remained in the mid 50's.
te relative size of the 3 people on the 60 ot wheel rim.
e wheel was burned and destroyed by ndals several years after this photo was cen.

Most of the early mine plants erected by W.G.M. have been demolished due to subsequent mining operations, but the old King Battery tailings wheel foundation still stands at this popular picnic spot just off the Kambalda Road (1km north of the White Hope mine, close to "Hunt's tree").

This project was not one of W.G.M.'s favourites as he noted that he did not enjoy taking instructions from the English engineer sent out by the Hampton Plains company;
"their engineer insisted that we build the plant upside down. Instead of building the battery at the top and letting gravity take the ore through the plant, they insisted the battery be at lake level so we had to lift the ore up to the cyanide vats."

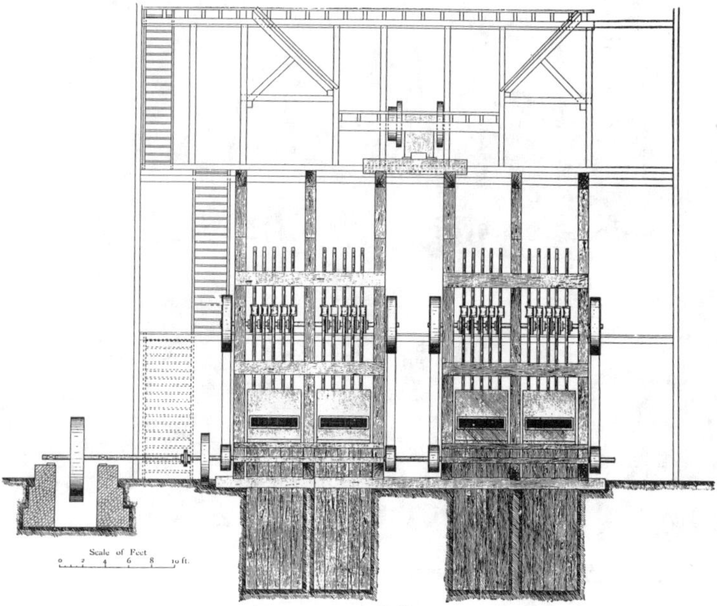

Fig. 68.

A typical quartz crushing battery as used on the Goldfields.

18

Batteries And Booze (1900 -)

While in partnership with Mr. Turrell, Mr. White, then Superintendent of Public Batteries, approached me with a view to designing a standard quartz crushing battery to be used throughout the State. This was really a splendid idea and would have been readily adopted by any sensibly minded business firm. But, I am afraid the government of the State is the antithesis of this. They just muddle along.

Seeing Mr. White later on the subject, he informed me that the Minister and Under Secretary for Mines were very suspicious, and if an idea were persisted in, they were liable to think one had an axe to grind. So the scheme fell through, the department bought up a lot of scrap iron plants of miscellaneous designs weights and ages and erected them at twice the cost that decent plants could be supplied.

If Mr. White's ideas had been adopted, one standard of spares could have been stored and, in supplying spare parts, could have been forwarded at a moment's notice, and many delays and needless expense avoided.

As an instance of this, at a State Battery forty miles from Leonora, a cam shaft broke. The plant was hung up and so were several prospectors waiting to have their crushings put through. Instead of being able to wire "Send new cam shaft." and get it delivered in a week, the manager got a trap and drove into Leonora then took the train to Kagoorlie and bought a cam shaft from the local merchant. He got this on the train and, after several days, it landed at Leonora. He got a special to take it out. It was duly landed on the mine, and a start made to put it in, when it was found to be a quarter of an inch too big in diameter. The manager, who was evidently not a born genius, had forgotten to take correct measurement and had bought a five and threequarter inch shaft when his cams and bearings were bored five and a half inch diameter.

The whole performance had to be repeated; the battery was hung up for about two months instead of a week. And that cam shaft probably cost the state ten times as much as it ought to have cost.

The State Batteries should have been a great blessing to the country and should have been the means of opening up and developing a lot of good new mines in the back blocks, and I know I am going to shock a great many people when I emphatically state that the reverse is the case.

If State Batteries had a fair chance, if they had been, as the saying is, on their own, in spite of government muddling, they would have done a lot of good and many outback towns would be flourishing now, producing much gold, instead of being deserted villages as they are now.

But they had a partner. This partner was akin to a man's drunken wife who drinks all his earnings and reduces his home to squalor and degradation.

The same Government that erected the State Batteries granted licences to publicans in that vicinity to sell intoxicating liquors. What good the batteries were doing was nullified by the pubs, very thoroughly. What exactly happens is this. Some prospectors have a mine with a payable lode, but have no money. They get credit at the stores for provisions, pending the getting out of a crushing. They may also get credit for a supply of beer and whisky from the publican. The storekeeper is regarded as a bit soft, but the publican is regarded as a good fellow. They get out a crushing, take it to the battery and have it treated, and get their gold which they sell to the bank. They pay their store account and their saloon account, and say, have £100 left over. Now this would make a nice start for a capital account to finance another crushing and ultimately equip the mine with machinery. No chance in life. They divide the money say £25 each, then adjourn to the pub, and there they stay on what is euphoniously termed "a howling spree" until they have either spent the whole lot or have been robbed of it while drunk. In many cases it is handed to the publican to "cut it out" or to take care of and he becomes vendor and trustee in one act. When he thinks they have been drunk long enough he tells them their money is done, and they had better get back to work. Then they arrange for further credit from the stores and the whole tragedy is repeated, again and again until their mine reaches water level and then the mine cannot be worked without machinery. They try to borrow money for machinery, but no one is fool enough to lend it. The upper levels are worked out, but one knows what is below, so capital cannot be raised to work it. They have nothing to sell so the mine is abandoned and lost to the State. It fills up with water, and who is going to prospect it? Thus many mines which under good management might have become big mines and produced thousands of ounces of gold, and employ hundreds of men, are now lying idle. Beside them is standing the State Battery rusting away in idleness, all poor and unfortunate victims of that soul destroying, nation destroying, imposter known as "Bung". And not only has Bung successfully prevented the development and opening up of new mines, and reduced the public batteries to a 10% working time, but he has on many occasions succeeded most effectually in shutting down big and promising mines after they have spent their capital and equipped the mine with up to date plant and machinery.

One illustration that has come within my own knowledge and can be verified will be sufficient to illustrate this sadly important fact.

We will call the mine the "Royal Banner Company". It was about twenty miles from a railway station. It had a lode eight feet wide that averaged twelve dwts per ton. This was a good mine, and had the makings of a big mine. It should be making a £1 per ton profit and employing 300 men. That meant supporting a population of about 1500 people. But what are the facts? A Company was floated with ample working capital. The mine was developed properly, equipped with a ten head stamp battery, an up to date suction gas engine and charcoal producer, concentrating machinery, and all accessories necessary to carry on successful and efficient mining.

There was a saloon in the vicinity, of course, and, what perhaps made matters worse, the beer after being carted twenty miles in open wagons under a hot sun, was not too good. But however bad it was, it was sold and drunk and, after each payday, the men would get so drunk through drinking this poison that they would not return to work under three or four days. And when they did, with continued drinking, many of them would be so muddled and inefficient that a reasonable amount of work could not be obtained which, in addition to accidents and mistakes, made conditions so bad that instead of a profit the mine was being worked at a loss.

So desperate was the situation and so unscrupulous was the publican that the management put up a very strong case to the Government, that the licence be cancelled and a new start made with the mine. Everything went well for about six months; the mine was working full time and making a profit.

The directors were considering increasing their plant and employing more men when, by back stair negotiations and possibly a little bribery, the publican's licence was restored and the old conditions renewed.

Being again faced with the familiar irregular running, the same old inefficiency, the same old financial loss, in short the same old "Bung", the company was compelled to shut down the mine. And so it is today, lying idle.

There are dozens of mines like this shut down for the same reason and although the criminal Bung is not always blamed for it, he is really the guilty person every time. As it is the last straw that breaks the camel's back, so it is the inefficiency and loss of time created through booze which, perhaps added to bad management, and extravagance in many cases, acts as the last straw in relation to the mine.

A significant remark was made to me one day by an old prospector named Walter Black. He had imposed on the military authorites, and had gone to the Great War. I am sure he must have been nearer fifty than forty five years of age when he left Australia, but he got to Gallipoli, had a go at the Turks, and received a bullet which he carries with him to this day. He was invalided home. The Repatriation Department assisted him on a prospecting tour in the Eastern Goldfields. At the time I met him he had not been successful in finding a mine.

He told me he was very anxious to find a good mine, one that would have the makings of a big one and, if successful he would apply for a publican's licence in the locality, then give the mine to anyone who would develop and equip it. By that means he would get all the profit from the mine and at least half of the wages paid to the men.

The old man was right. This is what happens in every case. The publican gets the lot. The misfortune being only that he kills the goose that lays the golden egg.

A brewer once gave me as his reason for putting a few hundred pounds into a mining venture, the selling of beer. He had no other interest in helping to develop the mining industry, his whole ambition was to sell beer. He knew he would get no dividends from the mine but, if it developed, well, he would get his money back many fold by the sale of beer. This same brewer told me that they did not allow their employees to drink beer on the premises. They required efficiency in their works (while striving to destroy it in all other industries). I sat beside this brewer watching Marie Correlli's *Holy Orders* being screened. This is a picture which shows the publican and brewer in their true characters. I enjoyed the picture but, oh lor, how I did enjoy seeing the brewer squirm and growl and grunt and condemn the picture as a lot of rot.

I was not surprised to read in the paper the next morning, "This picture will not be screened at Boulder City".

So the traffic goes on, killing, maiming people, destroying our industries and our legislators. And persons sanction it and make money from its existence, denying the people the right to vote it out on fair and square conditions. Thus vested interests override the interests of the people and the welfare of the nation.

W.G.M.'s brother, R.M. (Robert) Manners wedding photo with bride Carrie (nee Gathercole) at Broken Hill c.1895.

R.M. Manners as Senior Government Surveyor (1904-1932), returning with pack-camels over the Princess Range, five miles south of Wongawol (east of Wiluna W.A.).

Right

Home rebuilt 1909.

Left

Kalgoorlie home of
Mr & Mrs Charles Manners
(W.G.M.'s brother).
Demolished by cyclone 1907

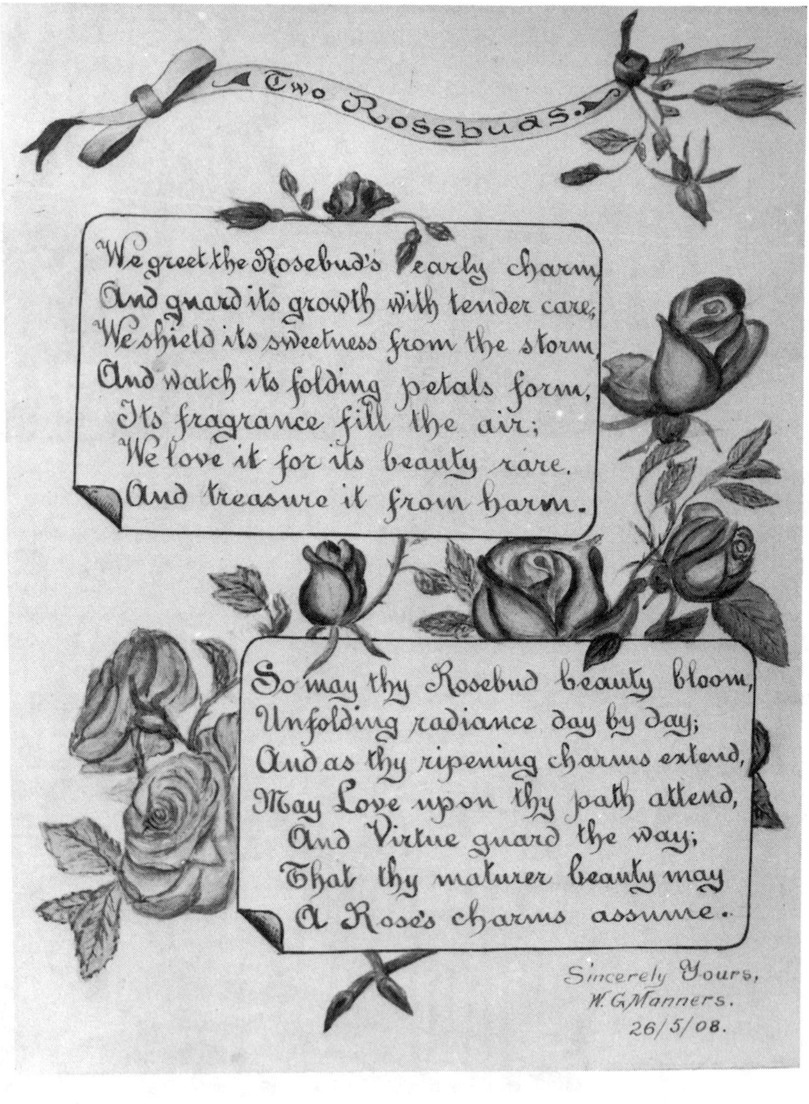

W.G.M.'s entry in the autograph book of the wife of Mr. Dorham "Dorrie" Doolette the joint-discoverer and promoter of the Bullfinch mining area.

● RIGHT: Killington's bakery in 1894 at the corner of Egan and Maritana-st Kalgoorlie—originally the post office block. Bread cost 1/ a loaf, water 1/ a gallon . . . Sent in by *C. Killington, Hillway, Nedlands.*

Kalgoorlie - Cnr Egan & Maritana St. 1894

Kalgoorlie - Cnr Egan & Maritana St. 1905.

PERHAPS THE MOTHER OF ALL BOOMS?

The sensational gold finds of the 1890's precipitated a boom in Western Australian mining shares.

In 1895 alone, £50,000,000 was subscribed for Western Australian mining companies, which were being floated in London at the rate of one per day, for 2 years.

Converted to 1991 dollars that one year's "raising" would be roughly equivalent to $11 billion.

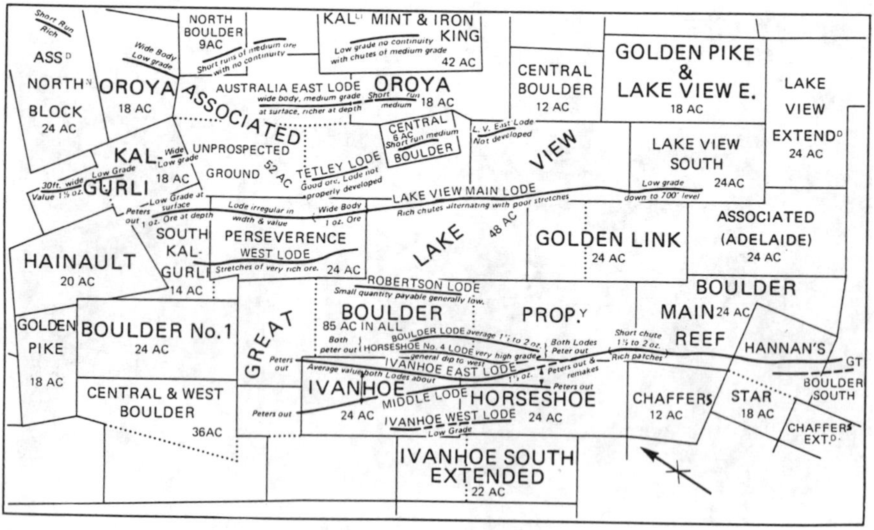

Kalgoorlie's Golden Mile Leases c.1900

19

Share Sharks' And How They Do It

As a mine-and industry-destroying institution, I award the palm to "Mr. Bung", he is pre-eminent, and easily the winner. But, although running a bad second, the share shark stands very high in the mine-destroying classes. As the brewer invests in mining for the purpose of selling beer, so the share shark invests in mining, not for the purpose of fostering the industry, but for the sole purpose of robbing the public.

I had bought shares and invested in mines, always on good developments or good information, but was always surprised to find that, in spite of the developments the shares always fell, and it was only by a lucky chance that one managed to buy or sell shares to advantage. I noticed that whenever the public was excited, and buying shares, then the bears got to work and always beat the public.

One day I was talking to a Jew on a visit from London, a member of the London Stock Exchange. He was complaining that the share market in London was practically dead. I asked him what was the cause of this. He said the provinces were not coming in, they were very shy.

"But", said I, "are there not enough people in London to make a market?"

"How can I play poker by myself?" was his significant reply.

The confidence-trick man cannot succeed without the innocent mug. The poker player who plays with marked cards cannot play poker with other players who use the same marked cards. But several poker players who play with marked cards can play very successfully with the honest fool who plays fair.

And this is what the share shark does, he plays with marked cards, and the principal marked card with which he plays is the system of "buying on cover". This ingenious method of robbing the public, I understand, was invented and introduced by that arch-scoundrel Horatio Bottomley, who seemed to be endowed with an uncanny faculty for devising schemes for inviting flies into his parlour. Of all his schemes this was the most ingenious, because it enabled the share shark to rob the public within the law, keep on robbing the public, and still keep out of jail. Bottomley was imprisoned for robbing the confiding public by means of his Victory Bond scheme. But the share sharks keep on robbing the public by their "buy on cover" scheme with perfect inpunity.

I propose to show how this swindle is worked so that those who read may know what to expect if they are tempted to walk into the parlour.

There is an open space in the City of London called The Bank. On the north side of this space stands the great Bank of England which gives the place its name. On the east side is the great London Stock Exchange.

These two institutions are the heart of the financial world so far as the British Empire is concerned. From this heart, financial arteries extend throughout the world. Every financial agitation at any point within this organism is immediately felt by the pulsations in this great heart. Inferences are drawn as to the cause of that agitation, and measures fomulated to take advantage of it or counteract it, as expediency may dictate.

The Stock Exchange is the pulse effected by any activity in the share market. Share sharks, becoming cognisant of this activity, are immediately on the alert and ready to invite the flies into their parlours.

We will suppose a certain mine, say in Kalgoorlie, has a capital of 100,000 shares controlled from London. The mine is rather under a cloud at present, just struggling along, but prospecting for a lode known to be in the vicinity. Say the shares are quoted at five shillings each. The men working in the cross-cut have discovered in the borings that they have struck the lode, and that the results are really good. They say nothing to the foreman but, having a few pounds saved up say about £70 each, they slip into town and order a thousand shares each at five shillings on cover, putting up a deposit of twenty five percent or one shilling and threepence per share.

An order coming on the market for 2000 shares sets the pulse going a trifle faster, and the sharks say "Hello! Something doing".

The miners, having secured all shares they can buy on cover, whisper the news to their friends and, next day, orders come home for 10,000 shares. This sets the pulse racing; the sharks cable the mine, but its officials know nothing. The sharks let them have all the shares they want "on cover". The mugs, hearing the whisper, buy all they can manage "on cover", putting up their last shilling as deposit.

Then the management becomes aware of the development, it is cabled home and made public. Then the public rush in, but the shares do not rise. The sharks will supply all demands "on cover"; they will sell 500,000 if asked for "on cover" when all the flies are in the parlour. There are still hundreds of thousands of shares for sale and the price begins to fall. There are sellers at four shillings, then at three shillings, then the brokers say to the unfortunate punter. "Your cover is exhausted, you must find another one shilling per share". The poor punter has put his last shilling on this "certainty" and cannot raise the necessary £50, so that broker has to "cover up" as it is called, that is, he sells the punter's shares and sends him a bill for the loss, say £30 or £40, which he has to pay by some means or other.

When all the blood is sucked out of all the flies that have come into the parlour, and the sharks have bought back at half price all the shares they sold, then the shares go up. They gradually are forced up to £1. In the meantime a new set of speculators come into the parlour, those who buy on a rising market. When they get loaded up with shares the bears get to work again and down they drop to five shillings; then the sharks who sold up to twenty shillings can buy back at five shillings.

In all this manoeuvering, scarcely any shares change hands. It is merely a book entry - and the balance is always in favour of the shark. Thus they are enabled by this "cover" system to sell half a million shares in a 100,000 share company, perhaps not possessing a single share to sell, and the proverbial fly that walks into the parlour has no more chance of getting out without having his blood sucked dry than has the actual fly that climbs the fabled winding stair.

I would not be bold enough to say that all sharebrokers are sharks. Indeed, I believe there are some very excellent and honest men among them but, being the exponents of a gambling profession, there is always the tendency and the temptation to take advantage of the mug when he comes along. Like betting, share dealing is reduced to such a gamble that it is a means of obtaining money without service and, in consequence, does not enrich the world but simply means that one person's gain is another's poverty. So, being such a doubtful profession, the one who makes the profit becomes hardened and does not allow himself to realise that he has robbed someone of his last shilling.

I had an experience of this sort myself, in which I played the part of the fly.

One day a friend came in and told me that there was a development on the Paringa Mine, that there would probably be a small rise in the shares. They had been quoted at one shilling and sixpence for some months. Regarding it as a good risk, I went across to a sharebroker named Hagerdart and ordered 500 shares. In a weak moment I forgot to put a limit on the price, and the broker, perhaps not often getting such a chance, did not suggest a limit. Imagine my surprise when next morning I got a bought note for 500 Paringas at six shillings per share. Knowing that I was the victim of a swindle I went to the broker and instructed him to sell. Next morning I got a sale note for 500 Paringas at one shilling and sixpence. They had never risen, and I had no redress. Of course the shares had never been bought or sold, and I had the pleasure of handing over just under £100 to square the ledger entry.

This broker is now a wealthy man in business in Sydney. A lady friend who had been visiting Sydney informed me with a note of admiration in her voice that he drove a Hudson Super Six. Of course that was easy, he had only to find half a dozen or so mugs like me in order to buy a Hudson Super Six. He probably finds hundreds of innocents or fools so honest that they regard all other men as immaculate also.

Before I close this chapter on "How they do it" I must refer to one other great financial anomaly that is foisted on the credulous public of Australia.

I am not imputing any dishonest motives to the promotors of the great Tattersalls' Sweeps, for I believe they are run on perfectly straight lines, and all the prize money that is available is honestly distributed. The fact is simply that the greatest proportion of the population are not endowed with the smallest faculty for figures or finance. This permits this great financial anomaly to exist from year to year.

If I were to ask ten people to give me five shillings each and draw lots as to who should win first prize of £1 and a second prize of ten shillings the people would laugh at me and ask if I mistook them for fools. Yet this is just what the promotors of Tattersalls Sweeps are doing, only the figures are so large and calculations so involved that they regard it as a good speculation.

They are so enamoured by the prospect of winning £5,000 for six shillings that they do not estimate the chances against them winning it and regard it as a good risk. Even when you tell them that in a sweep of 100,000 tickets that the odds against them winning the first prize are 99,999 to one they do not see the significance of the odds, the figures are too big for them. But this is not their main weakness in this financial anomaly. It is that very little more than half the money subscribed is distributed to the subscribers in prizes. I think it worth while trying to make this fact plain to those who are not adept at figures.

Let us take a sweep of 100,000 members at five shillings on the Melbourne Cup. Because the sweeps are run in Tasmania most of these tickets are sold through the post. The subscriber is asked to send six shillings for a five shillings ticket. Of this six shillings, sixpence is for mailing tickets and results, and sixpence is a tax made by the Tasmanian government on each ticket.

This leaves five shillings for distribution. Of this the promotors deduct ten percent for running the sweep. This amounts to sixpence. leaving four shillings and sixpence for distribution. Of this the Tasmanian Government take ten percent as an additional tax. This leaves four shillings and sixpence, that is, just over forty eight and a half pence. Then, of this amount, the Commonwealth Government annexes a tax of twelve and a half percent. This as as near as possible, sixpence, leaving three shillings and six and a half pence returned to the public for six shillings invested.

Let me put this into big figures:-

100,000 tickets at 6/-	£30,000
Less 100,000 sixpences postage	2,500
Leaving	27,500
Less 6d. per ticket tax	2,500
Leaving	25,000
Less 10% commission	2,500
Leaving	22,500
Less 10% Tasmanian tax	2,250
Leaving	20,250
Less 12½% Commonwealth tax	2,531
Leaving	£17,719

Thus, for £30,000 invested the public get back in prize money only £17,719. That is to say, in every sweep of 100,000 members at six shillings, the public loses £12,281. Or, in other words, in every two and a half sweeps the people lose the whole of the money invested in one of them.

Yet a mercenary government sanctions this imposition on the public for the sake of the revenue. And the more mercenary Commonwealth Government which administers laws in six states, five of which declare the sweeps illegal, and administers a Post Office through which such mail is not supposed to be carried, levy a tax on this gambling traffic, thus sanctioning it throughout the Commonwealth. This Government, which should protect the interests of the people who form their constituents, bow and stoop to robbing them like any common spieler would. And so all the people are fooled, all the time.

AUSTRALASIAN

Institute of Mining Engineers

This is to Certify

that Mr William G. Manners

Kalgoorlie,

Western Australia

is a Member of the Institute.

Elected 1909. D L Stirling Secretary

Troedel & Co

—THE—

Kalgoorlie Miner.

(PUBLISHED DAILY.)

THURSDAY, NOV. 18, 1909.

The membership list of the Australasian Institute of Mining Engineers is steadily mounting up. There have also been many additions to the Associate members' roll. The following West Australians were recently elected members:—Willoughby G. Bell, metallurgist, Kalgoorlie; Edgar A. Crutchett, mining surveyor, Boulder; James Drage, mechanical engineer, Boulder; John Dunstan, superintendent State batteries, Perth; Richard Eades, mine manager; Godfrey W. Ellis, mining surveyor, Coolgardie; Samuel Evans, mine manager, Fimiston; William G. Manners, mining engineer, Kalgoorlie; John McDermott, mine manager, Gwalia; James S. Paxton, diamond drilling engineer; E. Graham Price, mining engineer, Kalgoorlie; Harley B. Wright, metallurgist, Fimiston.

The General Manager of the Kalgurli Gold Mine, Mr. Frank Moss (3rd from left) entertaining the Colonial Secretary the Hon. G.H. Reid, Mrs. Reid and friends on the 100ft level. 28th January 1903.

Underground luncheon for visitors at the Kalgurli G.M. to mark the official opening of the Goldfields Water Supply 24-26 January 1903.

20

'Fields Water Scheme

Mr Nat Harper's Part
"Unnecessary" Pumping
Stations Criticised

Had it not been for a suggestion made by Mr Nat Harper to Lord Forrest in a speech at a banquet in Kanowna in November, 1895, there is little doubt we would not to-day have the great water scheme which has proved such a Godsend to these goldfields. Mr Harper was, at that time, managing the White Feather, Main Reef and Robinson mines.

In an interview, Mr Harper said that the great changes brought about by the water scheme would probably never have eventuated but for the late Lord Forrest's providential visit to Kanowna in 1895. Prior to that date, Sir John, as he then was, had visited the goldfields several times and, on one occasion stated that he had a scheme prepared, estimated to cost £6,000,000, but this was not within the realms of practical politics.

"I have searched the files in the Public Works Department to find that Sir John had given urgent and pressing instructions to the department to prepare a water scheme for his goldfields visit some time prior to his departure from Perth in 1895", said Mr Harper. "It would appear that many were propounded, but, for some inscrutable reason, were never placed before him, and Sir John had to face a famishing population without any scheme to put before them. The Goldfields League severely criticised his failures on the question and his retort was:- "You want the water; tell me how to get it, and if possible, I will get it for you". On the day prior to a banquet at Kanowna, he informed a Kalgoorlie deputation that he was hoping to meet someone with practical experience who could advise the Government how to solve the problem.

According to a report in *The West Australian* of November 27, 1895, Sir John Forrest, at this banquet, remarked:

"One of the greatest difficulties to be grappled with is that of water, and one of the chief objects of my visit is to find out how best the Government of the Colony can assist in the matter. The difficulties to be encountered with regard to the scarcity of water will, I believe, be overcome, although I cannot say how. What a transformation there would be if the district had a stream of fresh water running through it. How a supply can be obtained, I repeat, I cannot say, but it must be secured".

Sugestions at Banquet

"It was at this banquet at Kanowna, given in his honour, that I propounded the inauguration of the Goldfields water scheme to him. He came to that banquet as it were, distressed, and went away rejoicing. Lady Forrest

afterwards informed me that Sir John came home from Kanowna like a big schoolboy, bubbling over with delight at the idea of pumping water from the coast to the Goldfields."

To substantiate his claim to have convinced Sir John of the practicability of the scheme, Mr Harper produced letters in Sir John's handwriting and telegrams he had received from him. In one telegram the then Premier said: *"Many thanks for your letter. Have you any objection to my publishing, under your name, extracts from it? I feel it would do much good"*. Mr Harper gave the necessary permission and the letter was read by the then Minister for Mines in the Legislative Council in 1896. *"I am informed that it had a convincing effect in that council where Sir John had the greatest difficulty in getting the Goldfields Water Supply Bill through"*, said Mr Harper.

"That letter, written forty two years ago, stands unchanged to-day," he continued. In 1913 Lord Forrest was asked by a friend if the Goldfields water scheme was wholely and solely his. He replied: *"No, a man named Harper was the Father and I was the Mother of it!"* Mr John Woolcock, afterwards manager of the Golden Horseshoe and who was present at the banquet, said to me not long ago, *"I don't know whether you realise that your speech at that momentous banquet, when Sir John was convinced, was the most important ever voiced in Western Australia."*

"Subsequent to the banquet, Sir John, Woolcock, Jonathan Bray and myself had a discussion lasting till well after midnight, and Sir John, after sleeping on it, came to me the next morning and asked me to supply him with the quantities of water that I estimated would be required for battery purposes. I went into this and forwarded him the estimates and this formed the basis of the goldfields scheme." The engineer in chief, of the day was never enthusiastic about the water scheme, but yielded to the desire of his chief, Sir John, whose trouble was to get it sanctioned in Parliament. The scheme was described by members and others as that of a madman. Fortunately for Western Australia and the water scheme there was one practical man in the position of assistant engineer, in the early stages of the scheme at any rate. He had had experience in supervising the pumping of water in Tasmania. The early construction of the scheme for the first two years was slow and unsatisfactory; turmoil and friction existing right through the camp. In 1901 there was a grave danger of it being abandoned altogether, but Sir John came over from Melbourne to intervene and forced the issue.

In 1902 Mr C.S. Palmer took charge and the laying of the pipeline progressed at a rapid rate, the water reaching Kalgoorlie by the end of that year to the amazement and satisfaction of everybody.

Too Many Pumping Stations

Mr Harper was somewhat critical of the location of pumping stations, contending that four would have been sufficient. *"The No.1 station,"* he said, *"lifts the water 410 feet; No.2 station then lifts it 370 feet to Baker's Hill, 1080 feet above sea level. There a colossal and costly blunder was made in the original layout in that the water falls down from Baker's Hill to No.3 station, at Cunderdin, 792 feet above sea level, showing a loss of height of 288 feet and causing a serious economic waste in that the water has to be pumped up again, the cost of which, of course, falls on the consumers"*.

Mr Harper thought that, with the remodelling of mining plants, the scheme should also be remodelled and brought up to date. The last three or four of the pumping stations, he said, were only playing with the water. Kalgoorlie was only 1290 feet above sea level, yet the water was lifted 1993 feet which, when allowing for all considerations, meant an economic waste of at least 600 feet.

He quoted the lifts of the various pumping stations as follows:-

No.1 410 feet; No.2 370 feet; No.3 390 feet; No.4 396 feet; No.5 161 feet; No.6 187 feet; No.7 179 feet; and claimed that as stations Nos. 1 and 2 had already lifted the water 780 feet, or nearly two thirds of the natural elevation, so he felt many extra pumping stations were quite unnecessary.

Mr. Harper may have underestimated the friction losses in the pipeline but his early contribution to the project is noted. Everyone involved with this successful water scheme has a place in the history of our state.

It is also acknowledged, but not publicised that the water scheme completely paid for itself within the first five years of operation.

"Disappointment in the Swan River Colony followed the collapse of the gold rush to Halls Creek in the Kimberleys in 1886.

Following the granting of Responsible Government in 1890, the future of the colony lay in the hands of the pastoralists in the north, some gold mining in the Murchison and in the eastern "desert" where a modest gold mining enterprise was being carried out at Southern Cross.

On September 17 1892 the township of Southern Cross emptied following the news of a rich strike by Bayley and Ford at Fly Flat, (later to become Coolgardie). This first rich field drew prospectors from all over the world and more particularly from the Eastern States, to be followed by the strike by Paddy Hannan on June 17 1893 from which developed one of the richest goldfields of all times centred on the "Golden Mile" between the twin towns of Kalgoorlie and Boulder.

The gold strikes of the 1890s put Western Australia on the world map, but it was the Coolgardie Water Supply Scheme planned by the Civil Engineer, C.Y. O'Connor that ensured the success of the mining development and the prosperity of the State of Western Australia at the turn of the century."

R.M. Strickland in his introduction to the facsimile edition of the *Morning Herald* Guide Book #2 *Kookynie to Laverton.*

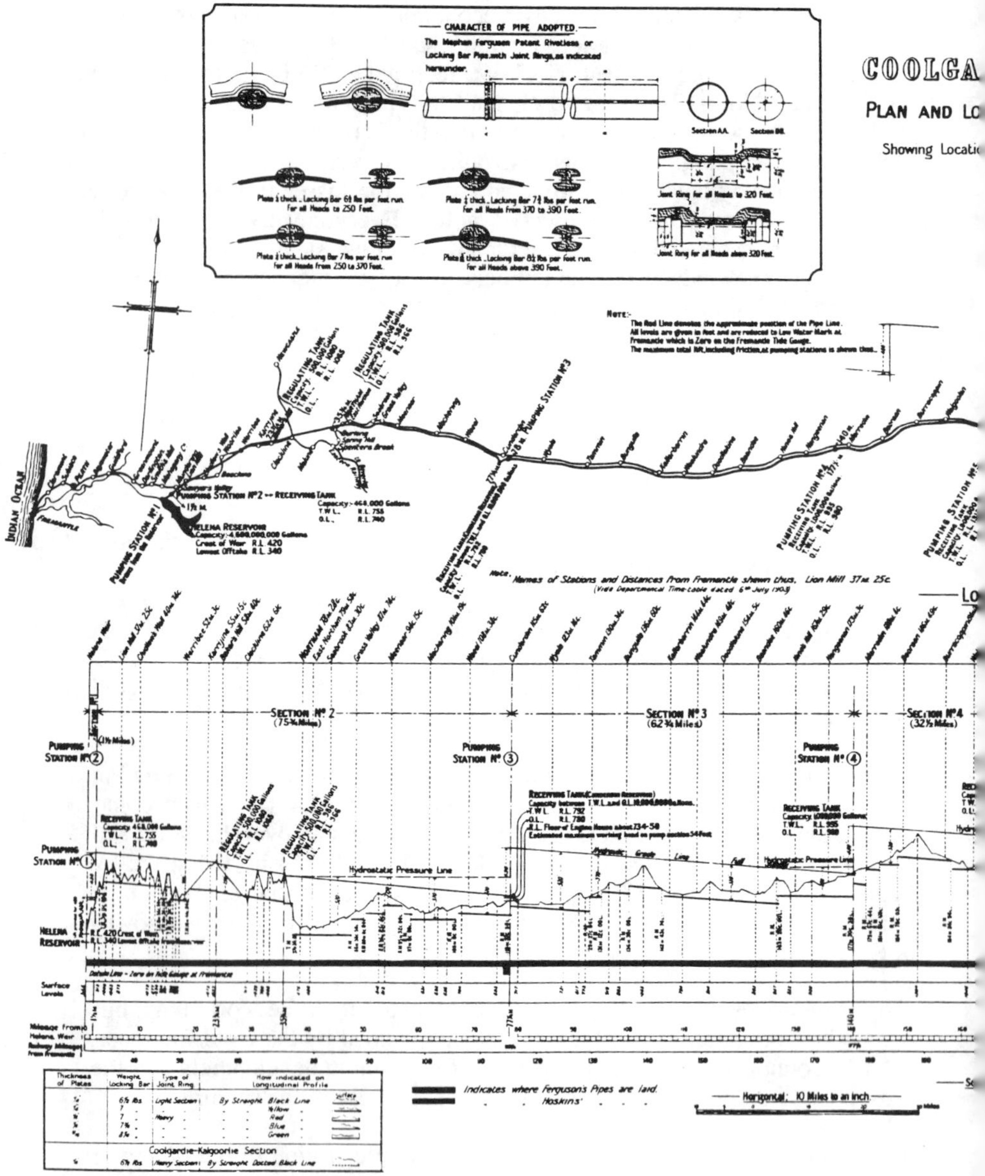
COOLGA
PLAN AND LO
Showing Locati
CHARACTER OF PIPE ADOPTED
The Mephan Ferguson Patent Rivetless or Locking Bar Pipe, with Joint Rings, as indicated hereunder.
Section AA.
Section BB.
Joint Ring for all Heads to 320 Feet.
Joint Ring for all Heads above 320 Feet.
INDIAN OCEAN
FREMANTLE
PUMPING STATION Nº 1
PUMPING STATION Nº 2 -- RECEIVING TANK
HELENA RESERVOIR
Crest of Weir R.L. 420
Lowest Offtake R.L. 340
REGULATING TANK
PUMPING STATION Nº 3
PUMPING STATION Nº 4
Note:- Names of Stations and Distances from Fremantle shown thus, Lion Mill 37M. 25C.
SECTION Nº 2 (75¾ Miles)
SECTION Nº 3 (62¾ Miles)
SECTION Nº 4 (32½ Miles)
PUMPING STATION Nº 2
PUMPING STATION Nº 3
PUMPING STATION Nº 4
Hydrostatic Pressure Line
HELENA RESERVOIR
Surface Levels
Mileage from Helena Weir
Railway Mileage from Fremantle
Indicates where Ferguson's Pipes are laid.
" " Hoskins' " " "
Horizontal: 10 Miles to an inch
Coolgardie-Kalgoorlie Section

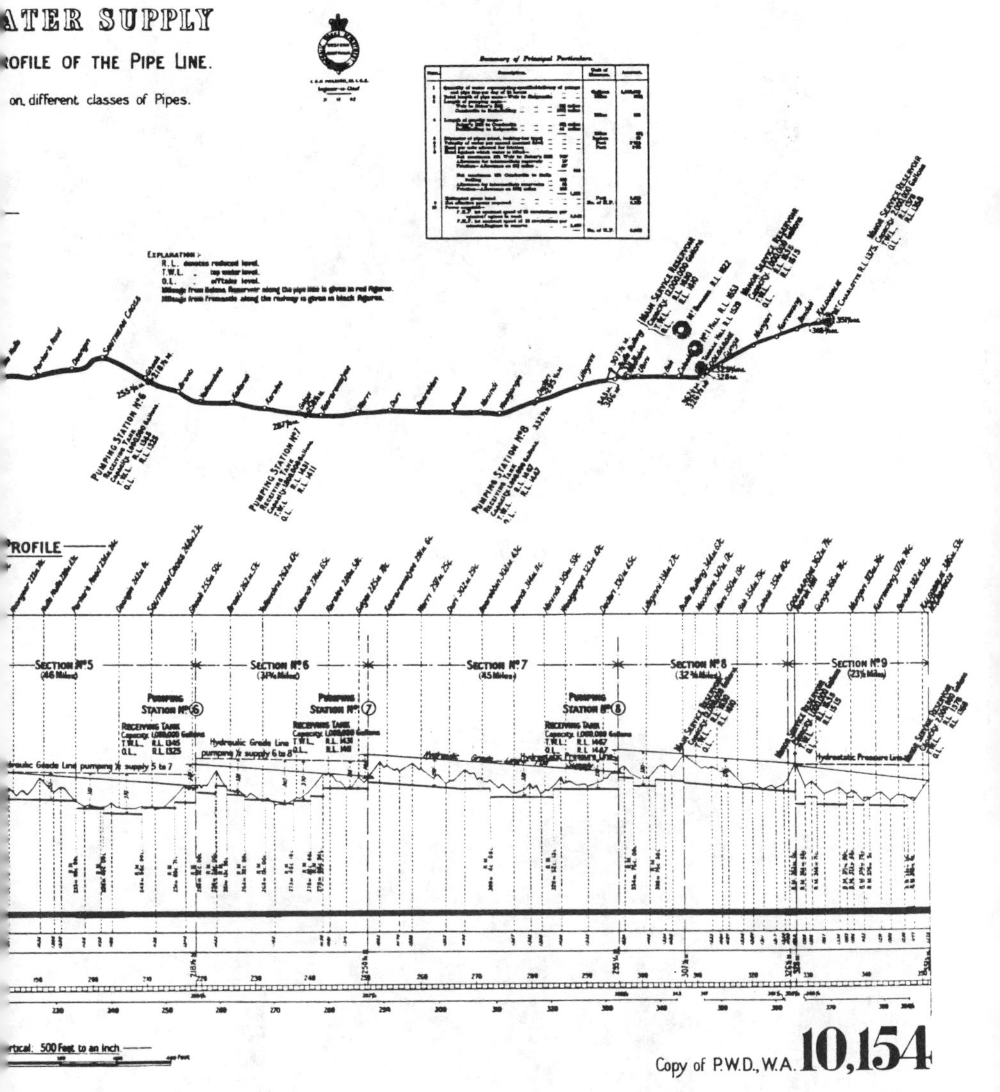
ATER SUPPLY
ROFILE OF THE PIPE LINE.
on different classes of Pipes.
EXPLANATION:-
R.L. denotes reduced level.
T.W.L. " top water level.
O.L. " offtake level.
PROFILE
SECTION No 5 (46 Miles)
SECTION No 6
SECTION No 7 (45 Miles)
SECTION No 8
SECTION No 9
PUMPING STATION No 6
PUMPING STATION No 7
PUMPING STATION No 8
Hydraulic Grade Line pumping ½ supply 6 to 8
Hydrostatic Pressure Line
Vertical: 500 Feet to an inch.
Copy of P.W.D., W.A. 10,154

Manners family turning the "first sod" at Denmark 1910.

The farmhouse - Denmark 1910

21

Go On The Land Young Man (1910)

It was in 1910 that James Mitchell, who was then Minister for Lands, travelled the State on a lecturing tour, his text being *Go on the land young man.*

Jimmy was a born optimist, painting a very glowing picture of the happiness to be achieved and the fortunes to be made by going on the land. I often wonder whether he really believed all he said in those lectures or whether he was just the professional politician and could not help it.

He said there were millions of acres of good land available, that the Government would provide it, lend the settler money to clear it and to fence it. They would also provide ploughs and horses to cultivate, and also seed wheat, and other things necessary to set the farmer up in life. In short, the settlers did not need money or anything else, all the Government would ask them to do was the work. All they had to do was to cultivate the land, the government would do the rest. This looked very tempting and very easy.

I reflected on our family history. My father had landed in Australia in 1853 when he might have bought half of Melbourne for an old song, and today we would have all been wealthy. He had lived fifty years in the country and, when he died, we had to buy a piece of land to bury him. That was all the land he ever owned.

Then here was I getting up in years with a family of five boys and, although I had owned land in the Silver and Goldfields, this had all deteriorated in value. I was practically going the same road as my father.

Yes Jimmy Mitchell, I would take your advice.

I resolved to take up a farm and develop it for some of my boys. So I journeyed to Perth, haunted the Lands Office, and worried the officials for some days.

I studied maps and was shown all the land that was available, but I could find none within forty miles of a railway line. I had travelled on the Bunbury line, the Eastern line, the Southern line, the Northern line, and I had observed millions of acres of land alongside these lines in its primeval state. Not a human habitation or a sheep visible for miles and miles on either side. I enquired about this land and was informed that it was all held in big estates by companies, lords and dukes and absentee landlords, and that it was not available for selection. It was being held until the population increased and became more valuable. Then it would be either cut up and sold or let to farmers at a price that would make it difficult for them to make it pay interest on the capital expenditure or, as in Old England, it would become big estates for aristocrats who wanted "room for their horses, equipage and hounds".

Thus was all the good usable land held out of use. If I wanted a farm I had to go away out into the back blocks where, if I grew a crop of wheat, it would have to be carted forty or fifty miles over bush tracks at a cost that would leave me in debt.

Of course the government might build a railway to the district some day, but there was no guarantee that this would happen in my life-time. So I gave up the quest for land in the wheat and sheep area, and was just a little doubtful of Jimmy Mitchell's veracity.

I heard there was good land in Denmark in the Great South West of Western Australia. The government had sold this land when it was a rich forest to Millars Karri and Jarrah Coy. at a low price. Now that the Company had taken millions of pounds worth of timber off it, the government bought it back at an enhanced price. —Dickens makes one of his characters say *The public is a hass.* I think he really meant *The Government is a hass.* It's simply marvellous how they do those things and keep on doing them.

Anybody but "a hass" would have let Millars have this timber on a royalty. The government would have received ten times as much as they sold the land for, and would not have had to buy it back.

Then consider the way they build railways into farming districts. First they sell all the land at from five shillings to ten shillings per acre on conditional purchase of twenty years. That means a rental of threepence or sixpence per acre for twenty years, then giving them the land.

They then borrow half a million pounds and build a railway to this district. This immediately increases the value of the land to £3 or £4 per acre and many of the Cockies who have paid only threepence per acre for 1000 acres sell out at £3 per acre then get back to the cities to spend it.

Now anybody but "a hass" would have built the railway first and sold the land afterwards and paid for the railway out of the increased value received for the land.

Of course I would not think of insinuating that some of the "hasses" friends were in the know and took up the land. That would be unkind, yet it is a fact that "hasses" sometimes get rich quick.

Well I liked the look of the land at Denmark. It looked like the Gippsland Country. I took up several blocks in the names of my family.

There was one block I was particularly drawn to. It had running water and looked very promising, although one could not see much of it because the scrub was so thick all over, and there was a regular forest of young Karri saplings on it about ten feet high.

The price was £14 per acre. I think this fascinated me too. I thought land valued at such a price must have some special qualities and be extra good. However, I learned later that this price was not owing to the special quality of the land, but to reimburse the government for its expenditure on "clearing this land and planting it with grass".

There were a large number of unemployed in Perth, (as there always will be while we have millions of acres deliberately held out of use), and the government had to get rid of them in some way or other.

Then Jimmy Mitchell had a brain-wave. He sent a few hundred of these unemployed to Denmark to clear some of the blocks ready for settlement. These men by the introduction of go slow tactics, so improved on the recognised "government stroke" that, by the time they had cut the scrub from the second acre, the first acre had grown up again to its original wildness. Thus, when they had finished clearing a block, one would never have known they had been there at all.

They broadcasted grass seeds on the unploughed surface. What was not washed into the creeks by the rains was either consumed by birds or, having taken root, choked by the more prolific growth of the scrub.

Then Jimmy had another brain-wave. It is astonishing what great ideas emanate from some brains. He purchased a few thousand goats and sent them into the district to eat the scrub. He erected dog-proof fences around a couple of blocks and enclosed some of the goats in these. Others he allowed to roam on the blocks that were supposed to be cleared. The dingos had a lovely time with these, eating many a banquet at Jimmy's expense.

The goats which were enclosed in the dog-proof fences died of starvation because the scrub that grew up was of too rank a nature even for goats.

All the costs of these brain-waves were charged up against the various blocks; the poor settlers were asked to pay for what was absolutely no use to them.

Oh Yes, Jimmy was a genius. He was just plain Jimmy then. He had not been knighted so it is not necessary to bend low, 'with bated breath and whispering humbleness', to say Sir James. In fact, I have heard it rumoured but I will not vouch for the accuracy of the statement that it was this great service rendered to the State that earned him his title K.C.M.G. meaning - Knight Commander of Many Goats.

This land that Jimmy cleared and "goated" cost me £20 per acre to clear again. And it took several years of agitation and protest to induce the government to write off this waste of money. They ultimately reduced the price of the land from £14 to thirty shillings per acre and, in spite of Jimmy's wonderful promises during his "Go on the land, young man" lecture it took two years' agitation to persuade the government to lend £10 per acre on the clearing.

Further reference will be made to this land in a later chapter when another great scheme of 'with bated breath and whispering humbleness' Sir James Mitchell will be discussed.

W.G.M. - The Farmer

Margaret - The Farmers Wife

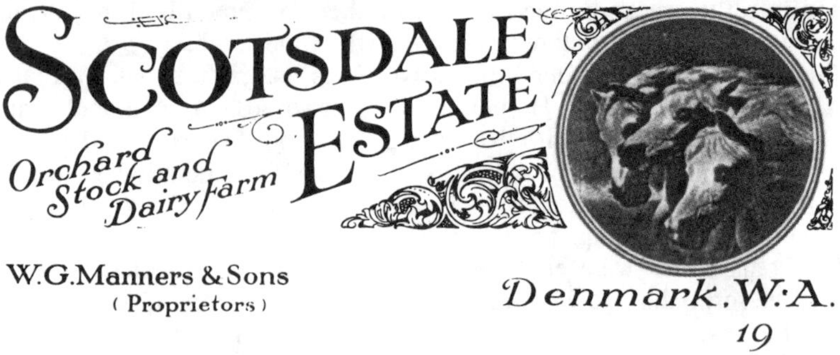

Two views of the farm shed (Margaret in R.H. foreground).

The farmhouse at Denmark.

Margaret with family on front steps.

Stone, Richardson, Orchard, Gerds, Kendal, Wathan, Hodge, Wolfe, Bourne, Dr. McKay, C. Dillon, B.L. Gardiner, J. McDermott, W.G. Manners, D. Head.

Daughter Muriel and Mrs. Margaret Manners enjoying the local transport in Durban c.1911.

22

More Mine Messers (1911 - 1912)

Shortly after this I received an offer to go to South Africa to design a large Mining Plant for a London company. Thinking that I would have an opportunity of doing something worth while, and using the vast experience I had acquired on the Broken Hill and Kalgoorlie fields, I gladly accepted it. Placing my business in the care of a friend, and putting my eldest son in charge of the farm at Denmark, I prepared for my departure.

It was just prior to the coronation of King George V, and a large crowd of Australian notables formed my ship companions. I had to go to London first to make the acquaintance of the directors and receive my instructions.

The trip over was uneventful and very enjoyable. Senator Varden of South Australia, and I soon became friendly, spending a good deal of time together. He had a sense of humour largely developed and could see the funny side of things. This is a saving grace.

Owing to the crowded boat he had been unable to get a deck chair at Adelaide, and the Deck Steward had promised him that he would get him one after we left Fremantle.

When we were about a few days out, three or four of us were sitting in our chairs on the deck when the senator came up and wanted to know whether he looked like an idiot or a silly fool. We assured him that although he did look a bit mad, it was not very serious. But he asked, "Really, do I look like a chump?" And we wanted to know why he should enquire. He explained that the Deck Steward had brought him a chair and he had given him half a sovereign, thinking that the Steward would pay the hire of the chair and keep the change. But the Purser had now approached him for the payment of five shillings for rent of the chair and, of course had had to pay and feel like a fool. The Steward, however, made a bad bargain because we all resolved not to tip him. Thus, for the five shillings he made by sharp practice he lost at least £5 in tips.

One day a gentleman had an audience to whom he was expounding the wonders of an infallible remedy he had for sea sickness. He had it made up in tablet form. All one had to do was to take one of these tablets three times a day and it would be quite impossible to be sea sick.

He was very convincing, and several of the passengers bought some of these valuable tablets. Among his customers was the parson.

The sea had been wonderfully calm up to this time, but the following Sunday morning it worked up a trifle rough. It was really nothing to write home about, and I would not have known it was rough, only our friend with the infallible remedy, and the parson and several of his adherents were too sick to appear on deck. There was no Church service that Sunday. I really think that the only passengers who were sick were those who had tested the infallible remedy.

At Colombo several Indian and Ceylonese Magnates came aboard to take part in the Royal festivities in London.

On arrival at Port Said we all went ashore to be accosted by all sorts of mendicants and pedlars. One man was displaying gold brooches and other trinkets which he offered us "Vera cheap - fifteen shillin". Senator Varden said, "Rot, I will give you eighteen pence for them". "All li, all li", said the pedlar and thrust them into the Senator's hand. He paid the shilling and sixpence for them. On close inspection found that they were similar to the jewellery one gets in a penny prize packet. However, it caused a lot of amusement, and the Senator admitted that he must be a bit silly.

Soon after my arrival in London I found that my hopes of using my ability to advantage were not to be realised.

The Company employed a firm of consulting engineers in London, and there was a consulting engineer on the Mine. The London firm showed me a letter from the Mine Engineer referring to my appointment which stipulated that I was to act entirely under his instructions. Not knowing what kind of a man he was, I was disappointed but hopeful. This was set back No.1

I was then shown a plan of a set of steel head-gear that had been designed by Mr Gordon Nixon, the Mine Consulting Engineer, for which a contract had been let to a firm for construction. They were sixty feet high and designed for a self-dumping skip.

Now my opinion was that a head-gear should be designed to accord with the rest of the plant, and in any case, should not be less than 100 feet high for so large a plant as was to be erected.

On examining the plan I discovered also that as they were designed, the skip could not possibly work in them. I pointed these things out to the London firm who had passed them and approved the contract, and was asked to say nothing to the directors. Then I was sent up to the Iron Works in Stockton on Tees to alter the skip arrangement and just try to work in the low head-gear. This was set-back No.2

There were wonderful preparations for the coronation ceremony: every available window and nook on the royal route was let at high prices; in available spaces large temporary galleries were erected; all the seats sold; and I was informed that folk would have to be in their seats at 5 o'clock in the morning or they may not be able to get through the crowds. This meant sitting there for several hours to see the procession pass. This did not appeal to me so I left London before the great coronation day, for South Africa.

It was an interesting sight while sailing down Southampton Harbour to see three lines of warships assembled, there were sixty in each line, all arranged in perfect geometric order.

At Madiera we took a trip up the mountain railway. This line is so steep that a cog rack is laid between the rails, and the engine has a gear wheel running in this rack. We had lunch at the top of the mountain then came down in cane chairs run on skids over a cobble stone road and guided by a man holding on and running behind.

We took a cab to the hotel where we had tea before returning to the ship. This cab consisted of a basket carriage on steel skids which was drawn over the cobble stone roads by two oxen. They travelled about two miles an hour and were driven by a man. A boy walked on in front calling out something. We suppose he was warning the public of the danger of getting in the way of this fast moving turn-out.

I spent a few weeks in Johannesburg and was struck by the size and extent of this wonderful mining field.

I visited the East Rand Proprietary Co., an amalgamation, I think of, three mines into one Company.

They employed 3,000 white men and 15,000 niggers. They had five miles of shaft and ninety miles of drives in the mine. Their output was enormous.

I visited the Moderfontein Mine twenty five miles north and the Randfontein, twenty five miles south of Johannesburg.

They have coal and gold in the same mine in one or two cases. The ore is free milling and consistent in values, so they have everything in their favour for large and cheap mining.

They were erecting a power house while I was there with 6,000 kilowatt generators and turbine compressors, with the idea of centralising the power.

They had magnificent machinery and were keeping it up to date by means of a Mines Trial Committee, who investigated every new idea that was suggested. If adopted, the inventor was recompensed and thus encouraged to use his brains.

Here I met Mr Pardiner a Metallurgist, who was going to the same Mine as I. He was appointed direct from Australia by Mr Nixon. We travelled to the mine together.

Although Mr Pardiner was receiving a much lower salary than I, he was appointed as Assistant Consulting Engineer. I was relegated to a position of Draughtsman. This was set back No.3 and was, in fact, the worst thing that could happen to Mr Nixon.

He was about the most ignorant man at mechanical engineering that it was ever my misfortune to be associated with. But, if he had had a little common sense, he would have made a friend of me instead of an enemy and I should have been able and very willing to help him and give him of my best.

He had designed and erected an experimental plant which was supposed to be on the point of starting up. It was designed on the lines of Kalgoorlie roasting practice. That is, dry crushing through Rock Breaker and Krupp Ball Mill, then roasting in Edwards' Roasting Furnace, mixing with water, fine grinding and amalgamating in grinding pans, agitating with cyanide and extraction of the liquid by filtering and precipitating the gold or zinc shavings.

I was sent down to the mine to see this plant and was told not to stay more than one day because Nixon wanted me at the office.

Because I had been sent out to study the ore and conditions at the mine, this order was tantamount to letting me know that I was not to have this privilege.

In reporting to Nixon, I suggested the plant seemed a trifle out of proportion inasmuch as the Ball Mill would treat thirty tons per day, the furnace about fifteen tons, the grinding pan about seven, the filter plant about 100 and that the plant was restricted by the one unit to seven tons per day.

He excused this on the plea that some of the machinery might form portion of the permanent plant.

I next asked when he expected the arrival of the Agitators. "The Agitators?", he replied, "By jove, I've forgotten to order them. How long will it take to get them made?" I suggested about three months, was rushed off to get out plans and specifications for them. This took a week, and another week to get in prices. The shortest contract offer was three months, because it would take another six weeks to erect them on the mine. Here was a five months delay through utter stupidity.

Nixon had designed and erected a ten head battery plant for another mine called the Straw Mine belonging to the same company. They had started crushing ore, but it was found that the costs for fuel were about four times above what it ought to have been.

The plant required about sixty horsepower to run it. Power costs should not be more than two shillings per horsepower day of twenty four hours, or £6 per day. As a matter of fact, the costs were in the vicinity of £24 per day.

While the Agitators were being made I was sent out to the Straw Mine to investigate and ascertain the cause of this loss.

When I got there I found the worst lay out of machinery that I had ever seen. They had six small boilers spread over a length of about 150 feet, and about 200 feet from the engines. They were what are known as dry-back boilers, all iron with a fire box in front.

The enormous surfaces exposed to the atmosphere, causing radiation of heat both in the boilers and steam pipes, would consume more fuel than would be necessary to run the whole plant.

There were two engines driving the plant, both of which I took indicator diagrams from.

There was a 200 horsepower horizontal compound engine driving the battery which was absorbing only twenty five horsepower. And there was a 100 horsepower horizontal compound engine driving the filter plant which was absorbing thirty horsepower. There was no condenser plant. Thus we had 300 horsepower in engines on a fifty five horsepower load and every engineer knows that reasonable efficiency was not possible under such circumstances.

Laymen may find this chapter rather uninteresting, but they will appreciate the extent of this anomaly when I say that it was almost equivalent to employing a draught horse to pull a perambulator; one can realise the waste of energy in such a case.

The only remedy I could suggest was to re-design the plant, scrap all the boilers, and put in one unit water tube boiler. But, in order to do this Mr Nixon would have to admit his mistakes. So there were no alterations made, with the result that the mine, not being a rich one, was run to a stand-still.

Thus are mines lost and ruined through ignorance and stupidity.

The Agitators being ready and built in, I was sent down to the big mine again to see that they were right. There was a hundred horsepower high class engine built in and, because the whole plant would absorb only seventy horsepower this was ample. So you can imagine my surprise when I found that they were building in a forty horsepower low class marine type engine to run the Agitators and filter plant.

This was repeating the error of the Straw Mine. But what was worse, with the limited tonnage possible, the crushing plant would require to be run for only four hours per day. And, as the furnace required only two and a half horsepower to run it, the 100 horsepower high class engine would be running for twenty hours per day with a load of two and a half horsepower, while the forty horsepower low class engine would be fully loaded all the time.

This was really making the Clydesdale push the perambulator.

When I pointed out this anomaly to Nixon he was surprised and asked how it could be remedied. I explained: by an intermediate shaft and pulleys connecting all the plant on to the big engine. On being told it would cost £50 he said, "We can't afford it now." meaning he could not admit his mistake; so the anomaly persisted and the fuel costs were three times too high. The plant was ultimately started but, of course, I was not allowed to be there. Mr Pardiner was put in charge.

However, one of the partners in the London firm of consulting engineers arrived and I explained how I was being treated. He insisted on my being sent down to the mine to study the treatment. I was sent down while he was in the district.

However, I was long enough there to observe that Pardiner was nearly as great a fool as Nixon. He knew only Kalgoorlie practice, and only a little of that, because he had been on back block mines.

I learned that the ore required a low temperature steady roast, that the roasted ore settled and set so quickly that vacuum filtering would be useless.

Nixon then wired for me to return to head office.

As the results obtained by Mr Pardiner were not very satisfactory, the directors appointed a Mr Allen who was highly educated and had a world wide experience, to make experiments and report.

He was treated by Nixon and Pardiner with the same hostility and antagonism as I was and got practically no help from them.

At this time one of the London directors was at head office and Pardiner, having failed to obtain a sweet roast, came to head office to try to get a sturtevant fan to blow up the fire and obtain a greater heat. I knew this was exactly the opposite of what was wanted.

The director sent for me and asked my opinion. I told him it was quite wrong and gave my reasons. He asked what I would suggest. I told him to have a second fire box built on the side of the furnace to get a low, well distributed heat as we would have in the big plant.

Pardiner admitted I was right. We were both sent down to the mine to make the alterations, but Pardiner was very angry and refused to shut down the plant until I threatened to wire the director. He gave in. When we got the furnace built in and again started, Pardiner got an attack of Malaria and went to the hospital for a few days, during which time Allen and I had a free run. We obtained a perfect roast from which we could get a 90% extraction by simple cyaniding without fine grinding. Our tests and assays were recorded when Mr Nixon appeared on the scene and asked how we were progressing. I told him very well and took him up to the assay office to show him the results.

He had a cousin who was working on night shift as fireman on the furnace and obtaining the best results and, unfortunately I mentioned his good work.

"Canningham," he said, "Yes! Yes! what shift is he on?"

I told him night shift. Nixon left that night to return to head office. Next morning Canningham had let the fire go down and the roast was rotten. On demanding an explanation he said it was an accident, but when the same thing happened next morning I saw it was a malicious conspiracy and explained to Allen.

That morning Pardiner returned, and although we had all the figures we wanted, we had no chance of a remedy. This bad roast destroyed the experiments for that month and thus Nixon, who should have been working in the interests of the company, was sacrificing its chance of success so that he may injure Allen and myself.

Because the vacuum filter was a failure, owing to the segregation of all the heavier particles at the bottom of the filter leaves, making it impossible to wash the upper part because of the extreme porosity of the lower portion, we had made a small model Burt Filter and found the even distribution of the particles consistent. We were able to get a practically perfect wash from a six inch thick cake and all our experiments indicated that with good roasting and Browns' Agitators and Burt Filters, Allen and I, with a free hand, of course, could have obtained a 90% extraction and made the mine pay a good profit.

However, Nixon and Pardiner, with the assistance of the London consulting engineers, had the majority of votes, so their plant was adopted.

I regret now that I undertook to design the plant for the company because I felt I was wasting their money and knew in my heart that their leaching plant and vacuum filters would never succeed.

The plant would cost nearly £200,000. It seemed like a crime to spend so much money on a plant on the advice of men who had proved themselves utterly incapable of constructing a plant that cost less than £1,000 without making a mess of it. But, so these London mining men are fooled, and Allen and I both realised that under Nixon's management even our design of plant would fail. However, as events turned out, Nixon never saw the plant. His sins had evidently found him out, but with the ingenuity that London boards display in the appointment of their Mine Messers, they succeeded in appointing a greater messer in his place. But this is a story for a later chapter, and will be told in its place. In the meantime I was packed off to London to design and purchase the new plant.

23

The Old Land At Work (1911 -1912)

Because I anticipated being in England for about twelve months (1911-12) I decided to take my wife, daughter and youngest son who was then about five years of age, with me. I cabled them to meet me at Capetown.

There was a little trouble at the start in London because the consulting engineers wanted me to work in their office and the mine company insisted on having me in theirs. I remained neutral and let them fight it out themselves, with the result that I worked in the mine office, and must say that, every facility was placed at my disposal to enable me to push on with the work.

Plans and specifications were soon completed and contracts let for the construction of the machinery. I found that this work was scattered all over England and Scotland. Most of the successful works make a speciality of one line. And they do their work well.

A big London firm of machinery merchants would probably get the contract for all the shop tools, such as lathes, planers, drillers etc. But, when it came to the time for inspecting these machines, one would have to go say to Birmingham to inspect the big lathe, to Cambridge to see the smaller one, to Manchester to see the planer, and to Bradford to see the driller. So it came about that I visited nearly all the manufacturing towns in England and Scotland and saw some wonderful performances in the way of turning out work.

For instance, in a shop where they make steam boiler economisers, which require a large number of cast iron pipes about four inches diameter and about eight foot long, they had two circular pits about fifteen feet in diameter, in each one of which they cast sixty pipes vertically at one time. The moulds for these 120 pipes were made in the morning before 11 o'clock. Between 11 and 12 o'clock the metal was poured into them. At 1 o'clock the pipes were removed and 120 new moulds set by 4 o'clock when another pouring of metal was completed by 5 o'clock.

Thus 240 pipes were cast each day. All other departments were on the same basis.

Nearly all this work was done on contract. The men worked very hard to make from £2 to £2/10/0 a week.

I saw one man making machine moulds for economiser scrapers; a shovel was his principal tool. He was stunted in growth, about five feet high, but sturdy. His grandfather and his father had worked on this job before him. He was working like a slave, at top speed all the time. I asked how much he would make a week at that work and was informed that he was an extra quick worker and made as much fifty shillings per week.

But I had to go to Glasgow to see the limit in slavery. Our buildings were to be all-steel constructions, I was inspecting one building in a Glasgow workshop. In passing from the office to the workshop I noticed some men rivetting on girders and I asked what wages they were paid. I was told the men were on piece work and earned as much as £2/10/0 a week per man, but they worked long hours for it. I asked what hours they did work and was informed that they started at 6 o'clock in the morning and had an hour for breakfast betwen eight and nine. They had from twelve to one for lunch, from five to six for tea then worked till nine at night.

There was a commercial boom on at the time. The workshops were all full of work and labour was scarce.

I remarked, "No wonder you are short of labour, these men do nothing but work, eat and sleep. If they go to Australia they would work eight hours a day, earn about £5 a week, play football or cricket or tennis on Saturday afternoon, and take their wives and families to the pictures about two nights a week - and would never come back to bonnie Scotland any more.

The original allocation of land in Britain is thought by some to have caused some problems that require new economic solutions.

If it is not possible to undo some of the problems caused by the original land grant favoritism of earlier kings, unrestricted competition will have to be the main saviour, to break the monopolies that impoverish the people.

The average wage at that time in Scotland was between £1 and twenty five shillings per week. A carriers' strike took place in Glasgow which threatened to shut down most of the industries. Their wages were twenty two shillings and sixpence per week, they were demanding twentyfour shillings. The masters ultimately gave it to them and settled the strike. They were too busy to risk the loss of the good trade that was offering.

These men mostly lived in flats, a whole family in one or two rooms, more like cattle than human beings. It was evident that there was something wrong with the economic system when these men and women who worked so hard and lived like cattle got so little return for their labour in the, at that time, richest country in the world. No wonder these men become socialists and communists. They think that it is the bosses who are grinding them down. They have no time to study the true principles of political economy. They listen to men as ignorant as themselves preaching communistic ideas and take it for gospel. They cry out for the nationalisation of the means of production, distribution and exchange. It is a fine sounding phrase, and seems to cover the whole ground, and they accept it as the whole policy of man. They cannot see the connection between their slavery and the land rents. They will not believe that the government granted private monoply of natural resources is the source of their bonds. When they realise that petition against the monopolistic land rents is the solution to their problem, then, and only then, will they obtain their freedom from this slavery and get a fair return for their labour.

In a large city where the commercial system is so complicated and involved, it is difficult for the uneducated man to see this fact because it is hidden behind what they see as "commercialism," "profiteering" and "slave driving". So poor old "Capital", that innocent and useful brother to "Labour", that friend and helper without whom labour is helpless, gets all the blame.

Many reformers think that drink is the cause of their economic slavery, but this is not quite so. As previously discussed, drink is a tremendous brake on industry, but in these large manufacturing towns it is more of a result than a cause.

I saw it in Glasgow at about its worst. These poor families working for a miserable pittance and living in one room have no comfort. Their homes are mere hovels. On the streets there is a saloon at each corner and one or two in between in each block. Inside is bright and cheerful and warm. What more natural than they should seek this comfort and drink and try to be happy?

I have seen in the streets of Glasgow thousands of children playing in the gutters at ten o'clock at night, and shivering women without hats or boots standing at pub doors, a miserable baby in their arms, waiting for the pub to close to take their drunken husbands home while, in the streets, the police were busy gathering up drunken men and women by the dozen.

Close up these saloons and you would certainly make these people better off and able to pay more rent, but you would deprive them of some of the comfort that makes their lives worth living. But, because their rent would be increased, they would be little better financially.

Some economists are now arguing that the fundamental cause of economic misery is the private ownership of land. "Nationalise" land rents and the people become prosperous, properly housed, properly educated. (Perhaps the pubs not then a necessity, can be abolished, thus confering another wonderful boon on society!).

It seems to me that this proposal may be ideal if starting from scratch, but with the problems already in place, that leaves competition to be used as an effective weapon to break the power of these monopolies. Free trade within and beyond national borders will prevent monopolies being used against the people. These lands would start to be drained of their productive workforce, causing land rents to drop dramatically in an effort to compete with these newer "lands of opportunity" across the ocean waters.

It appears to me that, unless these things are remedied from the bottom up, the British nation must go the way of ancient Greece and Rome.

The current monopoly of the natural resources of the country, protected by trade barriers means the concentration of all the wealth of the nation into a few hands, making the rich richer and the poor poorer. The marriage rate falls, the birth rate falls. There were a million and a half unmarried women in England while I was there. There were no husbands for them. The enterprising young men were already getting out to U.S.A., Canada, and Australia, and would never come back. The wages were so low that to marry meant poverty, no education, and slums for the children. And this in the richest country in the world.

Mr Lloyd George was making a desperate fight against almost overwhelming odds to remedy this state of affairs and I believe would have gone a very long way towards success, but the world war intervened and claimed his wonderful energy, ability and organising powers.

He introduced a bill to bring the valuation of land up to date and to impose a tax on unimproved values. To use his own words he was going to create "A discount on Dukes". The House of Lords threw out the bill. Then George brought in a bill to destroy the power of veto of the House of Lords, and threatened to create 500 Lords to carry the Bill through. This would have made Lords cheap. They signed their own death warrant. He had beaten the great house of obstruction.

Then he introduced his National Insurance Bill. This was strenuously opposed by the Great British Medical Association. They positively refused to go on his panels. George threatened to import doctors from all over the world and employ them as government servants. This brought the autocratic Medical

Association to its knees. The panels were filled inside a week.

What might not this man do? A little man of the people smashed two of the strongest autocratic institutuions in England. I honestly believe he would have saved the nation from its inevitable doom had not the war called on his powers. And, again, he smashed the greatest military organisation the world has ever seen. But this switched his energies from his great purpose. Now we must await the coming of the next great man who will be strong enough to smash his way through the formidable opposition of vested interests.

One thing The Great War did was to put Australia on the map so far as British education was concerned.

In 1913 I met many people who did not know where Australia was, or that it was an important part of the British Empire.

My youngest son attended a kindergarten school in Golders Green where all the children were brought to school by their nurses, much to the disgust of my boy. He would not allow his sister to take him to school for fear she would be mistaken for his nurse. This was, perhaps, the innate spirit of independence which characterised the Australian and impelled him to refuse to salute his officers, and to take the initiative when it meant winning a battle.

My boy came home one day and said.

" Mother, there are two foreigners in our school."

"Are there?" she replied "Who are they?"

"Little Don Petrio, from Spain, is one. and I am the other."

Of course these were only children, but they had doubtlessly insinuated this idea of "foreigners" from their parents.

On another occasion my daughter, who was about twenty years of age, was buying gloves in a shop in Golders Green. She was being served by a lady about ten years older than herself. My daughter remarked that these articles were dearer in London than in Australia.

"Do you come from Australia?" asked the lady.

"Yes, I am an Australian."

"Well I am surprised" she remarked, "You speak English very well."

Ye Gods! We, the proud descendants of the stalwart Britishers who pioneered the far off island continent of Australia, sons and daughters of the British Empire, and we speak English very well!

I learned another method of earning a living while in London. One day a very decent looking and well spoken young man came to my door with a basket of eggs. He told a good tale. He and his brother had started a poultry farm, having about 200 fowls at present. They wished to establish a round of customers to sell direct to the consumers, and they were selling their eggs at threepence per dozen less than shop prices. They could also supply chickens, dressed or undressed. My wife bought a dozen eggs and ordered a dressed chicken to be delivered on Saturday.

Some hours later while preparing dinner one of the eggs was broken. It exploded with a loud report and the smell was a thing not easily forgotten. The same result followed an operation on several others, with the result that the eggs were buried deep in the garden. Our embryo poultry farmer failed to materialise on Saturday with the chicken. I would be sorry for him if he struck one of his former customers a second time. But, in a place like London he could go a long time without having to repeat a visit to the same street.

The go slow policy was evident among wages men in London. Some repairs being made to a bridge which I had to cross every day. There was a plain brick wall eighteen inches thick being built and from the progress made from day to day, I should say that the workman was not laying more than 100 bricks a day. On such work he could have laid from 800 to 1000. These poor ignorant fools think they are making more work for their fellowmen and themselves, whereas, on the contrary, they are making commodities so dear that the workmen cannot buy. And, being a reduced buying market must of necessity reduce the demand and the amount of work available, thus creating unemployment and poverty.

Bill (1890), Charles (C.B.) (1894), W.G.M.(1863), Margaret (1863), Harley (1896), Stan (1898) & Muriel (Girlie) (1892). Photo taken 1905 at home, 143 Hanbury Street, Kalgoorlie.

Christmas Letter to 8 year old son George (born 1906)

Now Georgie my dear
Since Christmas is here
I must write you a letter today
To wish for you fun
Christmas pudding and bun
And lollies and presents and play.

I hope you'll recite
At the concerts at night
And show the boys what you can do
To make Stan look green
When he sees he has been
Knocked into a cocked hat by you.

If you have the good luck
To be eating roast duck
Or turkey or goose or a fowl
You can think of your Dad
In the bush very glad
Eating stewed kangaroo and boiled owl.

When Madgie and Bill
Have each had their fill
Of Christmas pudding and pie
And Nancy and Jean
Second helpings have seen
And think that a third they will try.

You can mention in jest
Castor oil is the best
To stop any after complaint
And Bevy and you
Could swallow some too
To show how delightful it ain't.

May you all be so very
Contented and merry
And happy and jolly and glad
That you'll always remember
This day in December
Is the wish of your loving
Old Dad.

W.G.M.
Hampton Plains
Dec.20 1914

24

Niggers And Num-Skulls (1912)

Having finished the plans and purchased all the machinery I returned to South Africa to complete the mining plant.

Mr Nixon had been dismissed. The London consulting engineers, through his bungling, had lost their job. Also, a new man with power of attorney had been appointed General Manager of the mine.

This was Jack MacMurty. He was called Jack Mack for short.

I believe this man was an expert underground timberman, but his knowledge of engineering, metallurgy and the science of mining was a negligible quantity. He attempted to hide his ignorance under an arrogant, bullying, swearing manner. I really think he was under the impression that he was camouflaging his ignorance, and bluffing educated men into the belief that he was an encyclopedia of knowledge.

He would walk round the plant with me every day cursing and swearing at everything, finding fault, and making out that everything was wrong.

He made one's life miserable, and discouraged confidence in his officers. If I would tell him anything that had happened he would contradict me. He got me to such a pass that I would tell him nothing. I really think the fellow meant no harm; his growl was quite natural and he did not know he was being offensive. He was one of those unfortunates who did not know how to use men and get the best out of them.

He said to me one day that he was going to put an air jet into the filter vat to keep the pulp in suspension and the "hydrogen" of the air would help to dissolve the gold. He did not explain how he was going to get hydrogen out of the air.

Pardiner was still in charge of the metallurgical department. Mack put him in control of the ore right through from the rock-breakers.

Because the ore was damp and moisture over one percent was fatal in ball milling, I had provided dryers. But Pardiner in his wisdom, neglected to use them, with the result that the finely ground ore would not gravitate from the ore bin to the furnace feeders, but banked up to form an arch over the bin door. It was found that, if force were used, such as striking the iron wall with a hammer, the ore would break away, avalanche down the shoot, then overflow the hopper. The top of this hopper, being about eight feet from the ground, would cause a cascade to spread over the land. But if the workman was careful and poked the ore with a rod until he got the shoot full it would run again for a short while.

This fact caused a very amusing incident much to the discomfort of Jack Mack. It was brought about by his bullying tactics.

He had been underground, then had been home to have a bath and change into a clean blue serge suit. He was making his growling round with his thumbs in his armholes and his chest well out.

When he came to the furnaces the ore was arched again and the furnace-man was trying to get it to run by gently poking it. Mack stood directly under the hopper and watched for a minute or so then said.

"What the B——— hell are you doing there?"

"The ore is arched, sir, and I am trying to get it to run again." was the polite reply.

Then Mack roared. "Hit it with the hammer".

"If I do", said the workman "it will ....". but he got no further. Mack was taking no back talk.

"Hit it with the hammer!" he roared.

The man ventured again;

"But sir ...."

"Damn your blanky eyes man, don't you hear me? I say hit the B—— G—— D—— thing with the hammer".

Bang! Bang! Bang! went the hammer with all the man's force. Down came the avalanche of dust, tons of it. It filled the hopper in an instant, fountained over the sides in a heavy shower. The next instant poor Mack was emerging twenty feet away from an enveloping cloud of dust that nearly choked him and changed the colour of his blue suit to a dirty grey. It sent him home a dirtier but wiser man, to have a bath, change his suit again, and send his blue serge to the dry cleaners.

On designing the push conveyors I had observed that conveyors with an eighteen inch stroke were not quite efficient owing to leaving a portion of the material delivered by the former blade behind.

To remedy this I lengthened the stroke and made the mistake of lengthening it to twenty four inches instead of twenty one inches. This had the effect of making the blades kick the heap of ore on the back stroke and throw up a small cloud of dust. This on a conveyor 100 feet long and four feet wide was a serious loss of ore and a great inconvenience to the workman.

The remedy was simple enough; viz. to reduce the stroke to twenty-one inches. I explained this to Mack and ordered new crank shafts. But at this time we had got to such a pass that he took to going past me and giving orders direct to the foreman under me, ignoring me altogether.

The conveyor was a hot ore machine for the purpose of cooling the red hot ore from the roasting furnaces.

I was astonished one morning to see the foreman carpenter with about twenty niggers completely covering the conveyor with oregon planking. I asked the carpenter who instructed him to do that and he said the manager had. I then went to Mack and told him the scheme would not only defeat the object of the conveyor, by preventing contact of the ore and air for cooling purposes, but that the timber would catch fire.

I think he really did not listen to me, but replied.

"We have to do something to keep down the dust, and that is how it is going to be done."

"Righto." I said, and left him.

It took the gang about three days to put the planks down. They used about £150 worth of oregon. About two hours after the plant had started up there was a wild cry of "fire". All hands were rushed onto the job to pull out the planking out, but not before it was all charred and destroyed, and the conveyor blocked up with wet ore from the water used to quench the fire.

I do not know what kind of fool Mack felt, but I knew what kind of fool everybody thought him.

There was another big bungle, this time it was Pardiner's ignorance that was to blame.

When designing the plant there was supposed to be a great scarcity of wood so the steam boiler plant was designed for Wankie Coal. But after the plant was delivered a forest was discovered some thirty miles from the mine, so a tramway was laid and a contract let for supply of wood for the roasting furnaces. On my arrival at the mine this wood was being delivered, there were already a few thousand tons stacked. It was still coming in and we were getting short of room to store it.

The manager remarked one day that he did not know where he was going to store it for the next month, the plant not being ready to start up for this time. I asked him how much more was to come, and was informed that the contract was made for 200 tons a day for three years. I asked who made that contract. It was on Pardiner's advice.

I then informed the manager that because we were treating 500 tons of ore per day, and the consumption of fuel in the furnaces would not be more than 12%, we would use only sixty tons per day. So, in the next three years, he would have to make provision for storing something like 130,000 tons of wood.

I do not know what arrangements the mine made with the contractor, but would imagine it would have to pay a large compensation because the contractor had been committed to a large capital expenditure on the strength of the contract. Of course, with these kinds of muddlers the mine had no hope of success.

Because the bullying and blustering was playing havoc with my health, I decided to get out. My nature was not adapted to such treatment as was meted out to me. There are natures which can stand this kind of thing, and I saw one on this mine. He was an underground foreman. I was never able to ascertain whether he had any intelligence; he never spoke except in monosyllables.

When he brought in his assay reports or such like, Mack would say.

"What the B—— H—— have you been doing here. Have you lost the B—— valves at the No.2 level?"

Sloane would say "Ya."

Then Mack would roar. "Well you'll have to —— —— —— find them again —— —— quick."

Sloane would say "Ya."

"You'll have to put more —— —— men on at the No.3 level." would order Mack.

Sloane would say "Ya."

One day Mack and I were walking along the No.1 level when we came across Sloane on his knees alongside the tramway points. He was busy sawing a piece of deal.

Mack roared.

"What the —— —— are you doing there? What the —— —— are you sawing that —— —— bit of pine for?"

Sloane said "Wedges."

Then Mack exploded. He roared and swore for about three minutes, finishing up with "That —— —— pine is no good for tram wedges, get a —— —— bit of jarrah, you —— fool."

During this eloquent peroration Sloane stopped sawing and looked at Mack much as a dog looks at his master when he reprimands him. When Mack had quite finished, Sloane quietly resumed his sawing just as if nothing had happened, and Mack, being evidently quite satisfied at having let off steam to show that he was the great "I am", passed on and left Sloane to his own devices.

Now if I had been a senseless stock of this sort I might have managed, but as it was I had to get out of it. However, having a three years' agreement, I could not afford to leave, so had to get Mack to dismiss me. This he did and thus enabled me to get a good substantial sum as compensation.

In the compound at this mine there were about 1,100 niggers. Not nearly all of them were employed, but it appeared that a large surplus of men had to be kept in the compound to meet all emergencies. As soon as these boys earned enough money they were anxious to get back to their native Kraals and buy a wife, or another wife, as the case might be. The compound was kept under very strict discipline. The Compound Manager was in full charge of all the niggers.

No man on the mine was allowed to strike or chastise a nigger. If he misbehaved or was disobedient he was sent to the Compound Manager. His case was investigated and a suitable punishment administered.

The punishment generally consisted of a certain number of strokes with a shambok. This instrument of torture is a strip of Rhinoceros hide four or five feet long. It is tanned hard for about two feet and the balance is pliable, thus forming a very effective whip.

The system is very cruel, but I am not prepared to argue that it is not the best method of maintaining dicipline with these simple children of nature.

I saw it applied one day and I felt disgusted at the pleasure the executioner took in the performance. And I was struck with admiration at the brave and unflinching way in which the nigger took his stripes.

The case was rather a peculiar one. In order to avoid collusion and conspiracy in the compound, the niggers are drawn from a number of tribes, many of which are feuding with each other. This prevents them coming together and planning a campaign against the whites which would be a disaster because it would mean, in this case, 1,100 niggers against thirty white men.

The culprit in this case was a new recruit from a remote tribe. There were only two niggers in the compound who could speak his language. He did not understand a word of English or Kitchen Kaffir, the language spoken between the blacks and whites, and it was evident that he had not understood his white boss's instructions. He appeared stupid. The boss was probably also a little impatient in sending him to the compound manager with a complaint of disobedience. The Compound Manager explained to me that the man was really innocent, but that he had to punish him to maintain discipline.

The Mine Manager's eldest son was a clerk in the compound. He begged for the privilege of being allowed to administer the sentence which was six strokes with the shambok.

The nigger was stretched on the floor on his stomach. Four niggers held his spread-eagled hands and feet. Young Mack laid six very effective strokes across the nigger's stern. The lad did not fool with the shambok, but laid it on good and hard and evidently enjoyed it very much.

The nigger laid like a log, his eyes with agony in them, on his tormentor. But he never winced or uttered a sound and, when it was over, got up and walked out of the office, not even putting his hands on the part injured, just walking off as though nothing had happened.

It is a point of honour with these men to bear pain and even torture without showing signs of fear or pain. I regard this as a characteristic of bravery that is beyond the power of white man.

I could just imagine what would happen to young Mack if someone were to administer to him the six strokes he gave with such pleasure. You would hear his howls in the next country and he would be rubbing the part for half a day.

I found the niggers a simple-minded, obedient, and good-hearted lot. I never found occasion to send one to the compound for punishment. It certainly required patience at times to make them understand, but I found that by treating them with a firm kindness and justice they were ideal servants.

My first acquaintance with my house boy was rather surprising. After travelling I had a number of dirty shirts, collars and underclothing. On reaching my kip, that is, a mud house with a thatch roof, I put these things out on my bed intending to make enquiries as to how I could get them washed. On arriving home at midday I was surprised to find them all missing. I thought they had been stolen, but said nothing until I had time to investigate the mystery. However, on arriving home again at 5 p.m. I found them all neatly laid out on my bunk, beautifully washed and ironed. The boy was all smiles when I complimented him on his good work.

I stayed in one boarding house where there were five or six nigger servants to attend to the bedrooms. I had no trouble with them, my room was always clean and tidy, and my bath and shaving water always promptly supplied.

But there was one man who had other notions as to how to manage niggers. His orders were always accompanied by a cuff or a kick and plenty of abuse. He simply treated them like dogs.

One day he took ill with a strange complaint. He was taken to hospital but, beyond it being some stomach trouble, the doctors could not diagnose it. He had probably a dose of ground glass or chopped hair in his stomach. These are favorite methods adopted by the niggers of getting their own back.

Glass or hair is mixed in the food in small quantities. It soon accumulates and sets up violent inflamation of the digestive organs.

Of course this cannot be proved against the niggers but it is a strange coincidence that men who abuse them generally fall ill of this strange complaint.

The niggers were also blamed a great deal for allegedly interferring with women and girls. Personally, I think it is remarkably to their credit that, where there are so many strong healthy young men and so few black women among them, there was so little of this class of crime. The men's wives are left back in the Kraal to cultivate their fields and look after their cows while the men come into the compounds for stated terms, say six to twelve months. There were a few black women kept in the compound, but the supply was totally inedaquate. That the natives themselves resent these crimes was evident in their assisting police to bring the criminals to justice.

A case occurred while I was there. A young lady school teacher was riding along a lonely road on her bicycle when a nigger accosted her, butted her off and, after a violent struggle, criminally assaulted her. The nigger was not known, but described by the girl. The whole police force, both black and white, were set on his tracks. He was discovered about a fortnight later in a Kraal some twenty five miles from the scene of the crime. His own countrymen had dealt out justice to him. They had performed an operation on him that would prevent him from ever committing such an offence again.

I do not know what happened to him after the British trial because I had left the country before he had sufficiently recovered from the operation to stand his trial. It is 100 to 1 odds that he was hanged.

This shows the native attitude to this class of crime. Their method of punishment has a great deal to recommend it to modern civilization; it does seem silly, when a man is found guilty of molesting a child, that he should be made a free boarder of the State for a few months and then turned on the world again, perhaps to repeat his abnormal practices.

In many cases niggers are severely punished for attempted assault on married and single women when, in my opinion, it is the women who should be severely spanked.

In one case a married women of good social standing had as a constant visitor a young man supposed to be her cousin. The nigger had seen familiarity between this "lady" and the young man. The natural inference was that if she was cheap to this man who was not her husband, she would be cheap to others. Perhaps if the nigger had been diplomatic and made his approaches gently and at an opportune moment they might not have been unwelcome. But his rush methods scared the lady and she screamed, thereby bringing assistance.

Another alleged lady went to have a bath one day and, having undressed, noticed she did not have a towel. She called the house boy and ordered him to bring one. In taking delivery she showed her bare arm and shoulder and breast. The nigger mistook it for an invitation, just as the average white man would. He tried to force his way into the bathroom, but her screams brought assistance. This women may not have been bad, but she was a fool.

Then again, when women come to South Africa they soon find that if they have good boys, they have nothing to do and become lazy. They have a cook boy, a house boy and a yard boy. These boys make the husband's breakfast before he goes to his work. Then the house boy brings the breakfast into the wife lying in bed. This may go on for a long time without trouble, but ultimately familiarity grows.

I know of one case where a very nice English girl came to Johannesburg to be married. They were a very happy couple. For two or three years no one would have thought ill of her. Her third child was a half caste; the family broke up.

I honestly believe that these South African niggers were an absolutely moral race until contaminated by the white civilization. And I believe now that, in the majority of cases of immorality practised in South Africa, the white women are to blame.

Physically they are fine specimens of men and women. They have a remarkable faculty for carrying things on their heads.

To the Kias on the mine it is necessary to carry water. The house boy brings it in a five gallon oil drum which he balances on his head without the support of his hands. This load will weigh over 50 lbs, or nearly half a hundred cwt.

When carrying the empty drum he will carry it on its side balanced on his head. It would be a difficult feat for a white man standing still, but the nigger walks along and looks side to side while the drum remains firm.

I met a fine strapping woman travelling on the road one day. She had three children running at her side and a baby in a sling on her back. She was eating a green mealie cob and carrying on her head a large tin dish which was piled about two feet high with pots and pans, mealie cobs, and other cooking utensils and articles of food.

As our motor horn sounded behind her she jumped aside and looked around. The miscellaneous pile on her head remained as firm as a rock. One of the children was carrying an old umbrella on its head and another had a tin of meat. They were being educated.

There seems to be one drawback where niggers are employed, that is a lack of mechanical efficiency.

I have seen a row of about 100 niggers digging up the field with mattocks preparatory to planting mealies. I should think one man with a double furrow plow and a decent team of horses would do as much in a couple of hours as this hundred niggers do in a day.

The nigger is supposed to be cheap, but the way he is employed he is, as a rule, rather expensive.

In sharpening hand-drills one white man usually does the job on his own. I have seen six niggers employed on the same job. One boy blew the bellows, one heated the drills, a third held them up behind, the fourth did the hammering, the fifth was a striker and the sixth carried the drill over to put it in the water.

I enquired why this was necessary. It was because they had to keep a large surplus of niggers in the compound and had to keep them employed. But the practice spoils the niggers for the reason that when one wants to get reasonable service from them they refuse to do it. They plead "Me striker boy", and he objects to blowing the bellows.

Thus it is that costs are not much lower in nigger countries than where white labour is employed exclusively and mechanical efficiency is at its highest standard.

Before leaving Rhodesia I would like to say a few words about the people there. Personally I never acquired that strong love for the country or felt the draw that Gertrude Page speaks of. It was not so bad in the mining districts where the people were mostly Australians, but in Salisbury where the seat of Government is as run by the Chartered Company, the people are mostly snobs.

They are principally government officials, younger sons of the English aristocracy who are sent out for colonial experience. The salary they receive from the B.S.A. Company is *£4* per week, paid monthly. They are often referred to as the "16-13-4 brigade". They mostly board, paying *£16* per month for lodging. They dress for "dinnah", keep a hack, and ride to hounds, with emphasis on the dipthong "hownds". Some of them have remittances, but the majority only have debts. One of the big stores there had no less than £25,000 on their books.

They have a very fine tennis club with *thirteen* ground courts made from ant hill clay which is very plentiful in the district. Some of the ant hills in Africa are as large as a four roomed cottage. But there is no sociability about the Club. A "16-13-4" snob would not lower his dignity by playing tennis with a common Australian engineer whose salary was only £1,000 a year. So although all members of the same club, each little clique played by themselves. A "16-13-4" snob could not play with a "person" who earned his living by trade.

Then the same thing prevails there as in other communities, that is intensive land monopoly. If anything, it is worse in Salisbury than in any Australian town.

In Salisbury there are about 3,000 whites. Yet the town is spread over an area of many square miles. There are large vacant blocks in the centre of the town which are many acres in extent. These are held by the big companies like the London and Rhodesian Mining and Land Company. There are several of these companies operating. They are holding this land to obtain higher prices. All with the cooperation of the government who have decided not to release any new land for building purposes, as this might compete with existing land and reduce its price.

If one wishes to build a house to accommodate his family he has to go about two miles from the centre of the town to get a quarter-acre block. Then he has to pay from £300 to £400 for it without any improvements. This makes a very dear home. The same land in any Australian city of a similar population would be worth only £20 or £30. This policy pertains throughout the whole of the territory with the result that the growth of Rhodesia will be retarded and will take perhaps a century longer to develop than it otherwise would. This will not matter to me because I am going back to good old Australia.

25

Aussie (1912 - 1915)

The railway men in the South African Union had gone on strike and it was likely that the railways would be shut down for some time. I was, in consequence, afraid to tackle the five days' journey to Capetown to board the boat to Australia for fear of being hung up in such a desert town as Mafeking for an indefinite period. I therefore pondered for a day or two on two schemes, one was to pay a visit to Victoria Falls, the other was to take train to Beira and get a Union Castle boat to Capetown.

I decided on the latter, but have since regretted that I was so near to the greatest water-falls in the world and missed the opportunity of seeing them.

There was nothing interesting on the railroad down to Beira, but some of the forest country was very beautiful and wild looking. Beira, like most of the Portugese towns, was about 100 years behind the times. The streets were mostly loose sand. The mode of transport is on narrow gauge tram tracks with trucks pushed along by men. The police, like most of the Portugese, are undersized in stature, but are quite up to the usual standard if estimated by their own opinion of their importance. However, when they run up against a burly British Tar it usually requires the whole available police force to get him safely under lock and key.

Such a case occurred while I was there. Two English sailors were having a night out among the pubs. One of them got a little too hilarious and was getting out of hand.

Two of the Portugese police attempted to arrest him, but, drunk as he was he soon had both of them sprawling on the pavement. They judiciously retired for reinforcements then soon returned with half a dozen recruits. The eight of them, after a severe struggle which lasted about fifteen minutes, and resulted in about half of the attackers being severely damaged, managed to overcome their victim and get him to the lock-up.

The next morning the sailor was brought before the magistrate, or whatever official dispenses justice in this country, and, because it was a serious case of violently resisting the police, the High Executioner decided to deal severely with the culprit. He fined him 10,000 reis or seven day's imprisonment.

This enormous fine quite appalled the victim. He had no idea how he was going to raise so much money; he had only about £5 in the world, and his boat was sailing the next day. He could not afford to lose his job and be stranded in a Godforsaken hole like Beira. However, he was greatly relieved when he ascertained that the equivalent of 10,000 reis in English money was about £2/16/0, so he paid the fine, and thought it was well worth the money.

There were very few passengers on the boat and the trip as far as Lorenzo Marks, another Portugese town, was quite uneventful.

We had no sooner put into port than a British Man-O-War steamed into the harbour and fired a salute of twenty one guns which was answered from the fort on shore. In a few minutes boats were lowered and 200 marines fully armed with rifles and bayonets were on their way to the wharf. We wondered what it all meant. When we went ashore we soon learned the cause.

It appeared that the railway strike was in full swing and on the border between Portugese East Africa and the Transvaal a lot of trucks of provisions were held up on the Portugese side. The Transvaal strikers had invaded the Portugese territory and raided the trucks. The Portugese authorities had cabled the British Man-O-War and, in less than twelve hours, it was in the harbour. The Tommies were sent up to the frontier, but the trouble was over so they had nothing to do in the way of fighting.

This incident, however, greatly impressed us with the power and far-reaching influence of that wonderful institution the British Navy.

By the time we reached Durban things had begun to move. General Botha had got busy and had called out the military forces which, with machine guns, had surrounded the Trades Hall and threatened to smash up the building with all in it if they did not surrender. There was no alternative so all the strike leaders were taken prisoner and smuggled off to some other town where they were imprisoned.

The rank and file of the strikers expected a trial in the morning and mustered their forces to try and effect a rescue. But nothing happened, and no one knew what really did happen until they learned that the twelve strike leaders were on a boat which had left Durban for England. In less than a week the railways were all in running order again although for some time they were under the guard of the military and police forces.

To many this may have seemed an autocratic and drastic step to take and, as a matter of fact, in Australia it would be so. But in South Africa the end quite justified the means. General Botha was a strong, level-headed man, a great statesman, and he foresaw what would be the result of a protracted struggle between the whites in that nigger country. You must remember that in Johannesburg, like all mining towns in that country, there are about ten times as many niggers as there are white people. Many of the niggers are somewhat educated and fully accustomed to the ways of the white. There are many Zulus among them — one of the finest black races in the world — the race of warriors which gave Britain so much fighting and expense to conquer, although it was only assagais and spears against machine guns.

There was therefore every prospect of the danger in the event of a quarrel between two classes of the whites that some black leader would arise, and taking advantage of the lion and the unicorn fighting for the crown, enter the fray like the little dog and knock them both down.

General Botha's action prevented this eventuality, and perhaps avoided a massacre of greater horror than any previously recorded in history.

We arrived at Capetown in due course and, after spending a very enjoyable week there took, the White Star boat for Albany. These boat trips were uneventful. The weather was good all the way and our time was spent in the usual way with deck games, bridge, concerts and dances.

So we arrived in Albany and were met on the wharf by several members of my family who were all very pleased with the reunion.

During my three years' sojourn in England and South Africa I had sent home something over £1,200 to develop the farm in Denmark and pay the college fees for my boys. I, in my charming optimism, expected to find some approximation to an equivalent value in development work on the farm.

This £1,200, and the expense of taking my family home, had left me with little more than £300 in hand, so you can imagine the shock I felt when I found that practically nothing had been done on the farm. All that money had been spent on travelling expenses to Perth, on deputations at my expense to induce the government to grant loans of £10 per acre to the farmers on clearing, and to supply cows on easy terms, and that a contract was let for clearing fifty acres on our place on a government loan of £500.

In addition to this, the school fees, £40 per quarter, had not been paid for a year, leaving £160 owing to the college. In addition to this, no one else had been paid. The balance of the debts to all sorts of merchants being about £440, made a total of £600 altogether.

This means that we had spent something like £1,800 in inducing the government to lend us £500. What kind of finance this is I do not know, but am inclined to think that it is 'damphool' finance.

Curiously enough I still had hopes of developing the farm even though we were starting with such a handicap. I realise now, ten years later, that I should have walked off the property then and there and taken the first loss. But, with a foolish bravery I recalled my boys from their colleges and set them to work on the farm. I took my eldest (Billy), who had shown such a remarkable penchant for political intrigue, off the farm with a view to getting him into some other business.

I could see this boy was naturally gifted as a politician; the thing was to get him into the political game. His faculty ought to make him treasurer of the state, for, do not all politicians do things upside-down? Do they not all spend £5 to gain two shillings and sixpence. In previous chapters I have spoken of this remarkable faculty. Do they not sell the land at a low rate, say two shillings and sixpence an acre in big areas, then borrow money to build railways and other public conveniences, then buy the land back at many pounds per acre for closer settlement purposes. Then the closed settlers get drunk and fail, and sell their land back to the big estate holders again for the purpose of selling it to the government, again for closer settlement. And so the dog goes on chasing its tail.

However, I made another mistake at this time. I tried to establish a commercial business at Albany. I bought a Ford car, on terms, to get the Ford Agency. There was a drought at the time, and later the war started and the business fell flat. My "politician" son took the car to Mt Barker and ran it for hire. This would have been a good move, but "politician" methods were introduced here also.

A party would hire the car to go for a joy ride, the charge being ten shillings. When they returned they would all be taken in to have a drink which would cost eight shillings, leaving two shillings to pay for the four shillings worth of petrol, and one shilling wear and tear.

On one occasion my son drove the publican and his family to Albany. This trip would be worth £2/10/0. I am not sure what money he got for it, but I know that he got some whisky and, on the way home, ran into a telegraph post, smashing the radiator and wind screen. These cost £13. Repairs to the telegraph pole cost another £2/10/0.

Another time he took a party out a few miles in the bush and forgot to close the oil tap in the engine crank casing. He came home with the metal run out of the connecting rods.

So things went on until I was absolutely broke and borrowed enough money from the bank to pay my fare back to Kalgoorlie, leaving my son and the car in Mt Barker. Of course he failed, but told me years after that he could have made a success of it if it had not been for whisky. Good old whisky. How many failures can be recorded to your malign influence?

C.B.M. Then about seventeen years old, attending college in Ballarat, recalled this experience:-

"Jack Blaikie, my cousin, had accompanied me by boat to Albany so we could spend the Christmas holidays at Denmark. I quickly realized the deplorable state of affairs at the farm and decided not to return to Ballarat. It was a big decision to make for it required at least another two years at the school to qualify, and I had my return ticket by boat to Melbourne.

"The deterioration on the farm during the twelve months that I was absent was most noticeable. Knowing that Dad was due to return during the following year I realized there was no alternative then for me to stay on the farm. Dad returned sooner than expected and received the biggest shock of his life. It must have proved the most tragic time in his whole life. I was at Denmark to welcome him home and know personally of his reactions. In my heart I cried with him."

Thus it was that instead of the country life I had so hopefully planned for myself and family, and many brilliant castles I had built in the air, I was compelled to sneak back to the goldfields which seemed the only place where I was capable of making a living. Yes, it was sneaking back. I had left there a man of some importance to go to South Africa to design a big mining plant which cost nearly £200,000. My salary was £1,000 a year plus all travelling expenses. After three years of this I came back to Kalgoorlie in a second-class carriage with only a few shillings in my pocket and a debt of over £400 hanging over my head. (I had paid off as much as I could).

Thus I roamed Kalgoorlie for a few weeks, then at last picked up a job to design a plant for a Hampton Plains company. This kept me busy for some time and afforded me some amusement. A man named James Speakman had been to Melbourne and raised a few thousand pounds capital to put a plant on this show then he came back. I was to go out to the mine with him starting at 9 a.m. the next morning.

He had met his mate who had prospected the show with him. They had had a night out to celebrate the great event of the flotation. They did not turn up at 9 or even 10 o'clock so I went off in search of them. I met them in the Victoria Hotel. Speakman was drunk and very loquacious, persuading Martin that he was the manager, that he was to have his own way and Martin and everybody else were to obey. After half an hour of this maudlin blather, and another drink, we set off to Clayton and Rintouls to get the sulky and spring cart. But the Oriental and the York Hotels intervened and delayed us so long with further assurances that Speakman was the boss etc. that we decided to have dinner before we started, arranging to meet at two o'clock. I was there but the others were not. I dug them out at the York Hotel and, after a good deal of trouble, got them to the stables while the horses were being harnessed. An adjournment was made to the

Kalgoorlie Hotel where further drinks and further assurances of Speakman's supremacy as a mining magnate were vouchsafed. At last we got a start about 4 o'clock on our twenty six mile journey.

On arrival at Boulder City Speakman had to call on his old friends the landlord and landlady of the Cornwall Hotel and stow away a couple more drinks, telling them of his importance as manager of Speakman's G.M. Co. This absorbed another half hour or so, then another stop was made at the Tasmania Hotel on the outskirts of Boulder City.

By the time we arrived at Lakeside it was dark. But, in spite of my angry protests, the manager would insist on having another at the Lakeside Hotel.

I had been driving in the buggy with Speakman, Martin and young Harry Speakman, a lad of about seventeen the manager's son, were in the spring cart. Because it was threatening rain I refused to go any further and said I would take the Woodline train back to town which would leave Lakeside about 7 o'clock. This action would have got Speakman into trouble so Martin persuaded me against it. He and I decided to continue the journey in the spring cart and leave the buggy for Speakman and his son.

The rain soon came on and we had a wretched journey. It was cold and wet, and to keep ourselves warm we walked most of the way. It took us five hours to do the twenty miles. We arrived like drowned rats about 11 o'clock. Mrs Martin soon got us some hot supper which was ready by the time we had changed our clothes.

We waited up until one o'clock in the morning, but there was no sign of Speakman. As a matter of fact, there was no sign of him for three days. It appears that after having a few more drinks and buying a bottle of whisky, he and Harry started off.

He was a good enough bushman. But whisky is guaranteed to eliminate all signs of efficiency, so they got lost. They had no food, only the Scotsman's luggage, and it rained for three days. They wandered through the bush all this time. No matches, no shelter and, although the whisky was consumed, they arrived back in Boulder City very wet, very hungry and very penitent.

They arrived at the mine on the fourth day and, although I was expecting a row for my revolt at Lakeside, there was no trouble. Jim was very humble and rather apologetic.

We got on very well after that and soon had the plant, a ten head battery with sixty horsepower Crossley Suction Gas Engine, running. It was interesting to notice how much Speakman didn't know about mining machinery. As for the mining part of it, he proved to be the usual fool optimist who opens up a face of orc, examines it with a magnifying glass and, where he can see gold, takes a sample, pans it off, then shows you a prospect of one ounce to the ton in the dish. Thus it was he estimated his values of ten dwts, per ton. But he had not taken into account the three tons of barren rock to every ton that carried ten dwts so when bulk crushings were put through the result was two and half dwts to the ton. Because this was just about one dwt below pay values the end was in sight; Speakman's greatness was like a man's who built his house upon the sand.

I left here to design a plant for the Ida H Mine[1] at Laverton, afterwards contracting to erect the plant. I had such good prices here that I made a good deal of money and paid off many of my Denmark debts.

During this time two of my boys (Harley and Stanley) went to the war to do

PROSPECTUS

OF THE

GOLDEN SPINIFEX MINING SYNDICATE, LIMITED.

(To be registered, under The Companies' Act, 1893, and Amendment Acts of Western Australia.)

CAPITAL - £2,750.

Divided into 550 Shares of £5 each.

300 Shares will be issued for subscription as follows:

Payable £2 on Application, and the balance by Calls at discretion of Di... but not to exceed £1 per share per month.

Provisional Directors:

ARTHUR FRANCIS, Esq., Kalgoorlie

C. D. McKENZIE, Esq., Kalgoorlie

F. LAUNDER, Esq., Kalgoorlie

Bankers:

WESTERN AUSTRALIAN BANK.

Solicitors:

ARNHEIM AND WATSON BROWN,

Hannan Street, Kalgoorlie.

Auditor:

(To be elected at first meeting of Shareholders)

Broker:

A. J. MELLOR,

Perth and Kalgoorlie.

Secretary *(pro tem.)*:

H. L. JERRARD.

THE Cpmpany is being formed for the purpose of *(inter alia)*:

(1.) Acquiring, working and developing Gold Mining Lease No. 24 acres known as the Golden Spinifex, and situate at Duketon on th Goldfield.

The gold mining lease prior to 1906 was worked but abandoned, excessive costs of carting, mining and battery charges.

The official record of crushings by the then owners, supplied by t is as follows:—

" Returns from Golden Spinifex Gold Mining Lease (Old No. by Mr. John Dwyer under No. 2035T:—Prior to 1906, 2642 tons for 1847

(2.) To acquire the mining plant, consisting of a 5-head batter water tube boiler, battery steam engine, jaw rock-breaker, head ge cages, winding ropes, buildings, fittings, and sundries, situate on Mac No. 20T, about 30 miles from the above lease. It is estimated the cos and re-erection of the same on the said Gold Mining Lease would not excee...

Form of A...

To the Directors of

The Golden Spinifex ...

I hereby make applica... named Syndicate and herewith ... per share, and I request yo... the terms of the Syndicate ... December, 1915, and I hereb... number that may be allotted to... holder of the said Shares.

Name in Full

Address

Description

Signature

The Golden Spin...

Received *this*

subject to the terms of the ...

the sum of

of Two Pounds per Share

above-named Company.

£ : :

for Shares.

ATE LIMITED.

Shares in the
que for £ be
hat number of Share
dated the 22nd
cept the same or ar
horise you to registe

Syndica

day o

spect

Particulars of Lease.

The Golden Spinifex Lease, No. 2035T, is traversed by two lines of lode running a few degrees east of north. The large reef is from 6 feet to 12 feet wide, and there are four shafts sunk on it, one only of which has reached water level. This reef samples from 5dwt. to 1 oz., and no doubt will pay to work later on. The smaller reef is from 2 feet to 6 feet wide, and has been stoped out to water level, viz., 30 feet for the full length of the pay shoot, 500 feet. This yielded 2642 tons for a return of £7489 over the plates and 12s. 6d. in the tailings.

There is an abundance of fresh water in the mine, but not more than can be handled by means of baling.

It is anticipated that the mining and treatment costs should not exceed 20s., and as there is every reason to expect at least 60s. values, there should be a profit of 40s. per ton.

The reef is a true fissure, and is almost certain to live to a great depth.

General Remarks.

It is anticipated that there will be no difficulty in obtaining a subsidy from the Government if required.

In 1904 the State Mining Engineer reported on the Erlston and Duketon districts, and the following extracts are from his report :—

"The township of Duketon is on the Golden Spinifex line of lode which runs a long distance through several leases. There appears to be three or more parallel reefs forming a run of lodes. Most of the workings were from prospecting shafts 60 to 80 feet deep, down to the water level. The water level comes nearer to the surface going northward, owing to the fall in the ground, being at about 80 feet at the south end and at 10 feet in the Morialta at the north end.

"The reefs are from two to six feet wide, and have the smoothed walls characteristic of fissure veins. Very fair crushings have been obtained from several of the leases, and the line of lease appears to be gold bearing throughout its length. The Golden Spinifex line of reef seems a very promising one."

Application for shares should be made on the accompanying form and forwarded to the Secretary of the Company, 11 Exchange Buildings, Kalgoorlie, together with the amount payable on application. If no allotment is made the deposit will be returned without deduction, and where the number allotted is less than the number applied for the surplus will be returned. The Company shall be deemed to be formed and the shares will be allotted on 200 shares being applied for.

The following contracts have been made, viz. :—

(1.) Agreement dated the 11th November, 1915, between John Dwyer, of Duketon, of the one part, and William George Manners, of the other part.

(2) A contract dated the 22nd day of December, 1915, made between John Dwyer of Duketon, and H. L. Jerrard of Kalgoorlie, accountant, for and on behalf of the Company. Under this contract John Dwyer is to receive 250 fully paid-up shares in full consideration of the said Gold Mining Lease, and the sum of £2,000 for the Battery Machinery and Plant. This sum of £2,000 is payable out of profits.

Forms of application for shares are attached.

Copies of the Memorandum and Articles of Association of the Company and of the contracts above mentioned may be inspected at the office of the Company's solicitors, Arnheim and Watson Brown, Park Buildings, Hannan Street, Kalgoorlie.

Dated this 22nd day of December, 1915.

their duty to their country, and the third (Charlie) went a little later, as soon as he had reached the required age of eighteen. It was noble of them to go. Although we were very anxious about them because they were all in the firing line, we were very proud of them; it was good of all those brave boys to offer their lives for their country and make room for a lot of slackers to take their jobs and make themselves very snug and comfortable at home.

It was while I was at the Ida H that my "evil angel" visited me and tempted me to speculate in the Golden Spinifex Mine. This was a mine owned by an old friend of mine, one Jack Dwyer. They had a lode three feet wide worth one ounce to the ton. Jack was an optimist who inspired me with his optimism. We bought the Mulga Queen Battery plant for £200 then came to Kalgoorlie, floated a syndicate, and got to work. We bought a pump and put it in. I went up to see the mine, but at the time of my visit I was not able to get in to see the lode because the drives were full of silt from the wall of the lode where it had been cut. This should have warned me off, but Jack's optimism beat me again. So I let a contract for fifty feet of driving. When this was nearly accomplished I called the Mines Inspector to go up with me to pass the work.

In going below we found they had cut the lode, but the drainage was too severe from the wall, so they drove on twenty feet further and cut it again with the same result. They then drove on ten feet further and cut it again. When I got there they had broken a hole about two feet in diameter through the lode and were lathing it up in order to prevent the hanging wall from coming through. The fact was that there were two "porridge" walls.

I asked the contractors how they were going to get the lode out and put in their timber set.

"Oh" they replied, "that is easy enough."

"Yes, I know it is quite easy, but I would like to hear from you how you propose to do it."

All I could get from them was that it was quite simple.

I saw then that they had been stringing along my good friend Jack Dwyer and had done thirty feet of driving after they discovered the thing could not be stoped. They knew that I saw through them and, that night when I went down to their camp to give them the sack, they told me that they were pulling out.

I told them I knew they were because I had come to push them out as a pair of imposters. That was the end of the Golden Spinifex and I was again on my beam end, having lost several months fooling with it and incurring liabilities amounting to about £250. I had to start all over again.

Fortunately I obtained a job to erect a set of Poppet Legs at the Riverina South Mine at Riverina. This led to much designing work and further contracts. The job pulled me out of my present difficulties and enabled me to start again my business in Kalgoorlie. But business was not very brisk; there was no construction work going on and I degenerated into a dealer in second-hand machinery which kept the pot boiling between jobs.

So the time passed during the progress of the war which was always very much in doubt. Two of the boys (Harley and Charlie) got wounded; Charlie (C.B.M.) very badly and was in hospital for some months. Harley subsequently went back to France and was unfortunate enough to be gassed.

These boys all going to the war also meant the eldest going back on to the farm. Billy was married and had several children, so it was not expedient for him to go to the war.

He specialised in pigs, having as many as 200 on the farm. But as he had to buy all his feed, because there was no market for pigs, the farm landed in further difficulties and increased its liabilities. I tried to induce him to dispose of the pigs and plant potatoes, which were ranging from £16 to £30 per ton, but my persuasive powers were not sufficient to overcome his pig optimism. So we continued all through the war, while all other farmers were making their fortunes, to get further into debt.

Some folk seem to have a wonderful facility for doing the wrong thing every time. My lad seemed to have this wonderfully developed.

At last came the joyful news of the Armistice, the defeat of the Huns, the subjection of the Kaiser. Oh, what joy prevailed! What excitement, what cheering, what laughing, what tears of joy were shed. There were many sad hearts, rejoicing at the victory and mourning the loss of their dear brave sons. There were the more lucky ones, like ourselves, who wept tears of joy and thankfulness that peace was to be declared and our three boys were still safe and would return.

What a sacrifice it is to give your sons and lose them, even for your country's good. But what a joy to have given them, to know they had helped towards victory, were still safe and coming home. Coming home. Oh, that home coming! What a prospect. What a welcome they would have. How we would all rejoice to meet them, to feel proud of them and know they were men who could always hold their heads up for having done their duty, answered the call of their country and helped to save it from the yoke of oppression and slavery that Germany would have assuredly imposed on us had there not been sufficient loyal and brave boys to uphold the honour of the flag that has never yet been humiliated.

Being in this proud position makes one feel a contempt for the slackers who failed to answer the call, who hid behind women's peticoats and pretended ailments because they were too cowardly to go. I can now understand the feelings of the old Scotsman who went to hear the Kennedy family of Scottish singers who toured Australia in the Eighties. He saw the minister there, the Rev. J.W. Inglis. He met him next day in the street and rushed up to him.

"Ah, Minister Inglis, I saw you at the concert last night."

Because it was not considered proper for a Presbyterian Minister to visit a theatre, Mr Inglis thought he was in for a reprimand and replied.

"Well, was there any harm in that?"

"Na, Na, man." replied the Scotty, "But tell me noo, wasno it grand?"

"Aye aye it was real fine."

Then Scotty enthused.

"Losh man, when auld Kennedy came out and sang the "Lang 'o the leal", man, I fairly wept. But when he came oot and sang 'Scots wa' hae' Lard bless you I felt ma auld Scotch blood bile and there was an English cheil sitting next me an' I felt inclined to gae him a cloot the side of the lug."

That is how I often felt when I saw these slackers in their nice snug billets enjoying the luxuries and pleasures of good salaries and good homes while those who had sacrificed everything for their country had missed all these chances of education and professional experience.

But such is life, and I feel sure that those who missed the chance of being men and true Britishers will always feel ashamed when they look a dinkum Aussie Digger in the face. And the monuments that meet the eye in every town and village, erected in honour of those who made the supreme sacrifice, must make them feel that they have missed something that will make them stammer and blush when their children say, "And did you fight in the Great World War, Daddy?" And what if that son should be ashamed to hear the "No, I was too wise for that."

Note 1.

IDA H MINE AT LAVERTON

"Incidentally I went out to the Ida H on Saturday last. This is where I first met your grandfather W.G. Manners in 1914.

I was helping with all the surveys and W.G.M. had the engineering design and installation.

This was my first visit since August 1915 (when I left for World War I).

My visit was in connection with a project to extend to the surface an incline shaft which operated at 500ft. down to 1300ft."

Letter to Ron Manners from G. Spencer Compton

February 2, 1968.

THE

Kalgoorlie Miner

TUESDAY, JULY 30, 1918.

WOUNDED.

Mr. and Mrs. W. G. Manners, 13 Victoria-street, Kalgoorlie, were notified by the Defence Department last Saturday that their second son, Pte. C. B. Manners, had been wounded in France. No other particulars were given. Pte. Manners has served for about 15 months with the 44th Battalion, A.I.F.

26

The Riot (1919)

Although conscription was turned down on two occasions, mainly by the influence of the Labor Party assisted by the Roman Catholic and other pro-German parties, most of the loyal and patriotic men had enlisted by this time, leaving the slackers behind to vote on the question.

Employers, including the Government Municipal Councils and Roads Boards, had promised that those who enlisted would find their jobs waiting for them on their return from the war. This time had come, the men were being repatriated, and it was up to employers to honourably keep their promises. They had not only promised this, but had guaranteed that, in the event of any vacancy occuring, preference would be given to returned soldiers.

The government had honoured this pledge by taking the men back in their old jobs but, possibly for fear of creating trouble of losing votes, they kept the slackers on as well and thus got outrageously over-manned, as the story of loading the boiler will illustrate.

On the mines and other works which must be operated on commercial principles, many of the men who had usurped soldiers' jobs when they enlisted were dismissed and the soldiers put back in their old places. Besides, this preference for soldiers policy, was running counter to the old preference to Unionists cries which had been so stubbornly fought for, with indifferent success during the last forty years.

Again the Labor Party had been split in two by the conscription campaigns; it was divided into two parties who named themselves the National Labor Party and the official Labor Party. Nearly all the loyalists left the official Labor Party or were forced out. Such sterling Laborites as W.M. Hughes, Spence, Pierce, Watson, Scadden, Cann, and many others constituting the actual brains and sinews of the party left their comrades for the nobler paths of honour and country and freedom.

(Sketch by W.G.M.)

This left the Roman Catholic section in a dominant position; 75% of the offices of unions were filled with R.C's and pro-German members of the union.

When the official Laborites found that the policy of preference to soldiers was a real thing they began to make trouble. The soldiers found that it was sufficient for them to belong to the Returned Soldiers and Sailors Association. A large institute building had been erected ready for their return, by the patriotic people. This was their head-quarters, well organised and well staffed and officered.

The Official Labor crowd know of only one method of gaining a point so after some meetings, they organised a mob of roughs who raided the mines. All those who could not show union tickets were ill-treated, some threatened with death such as throwing them down their shafts. In fact, one man named Edwards was so maltreated by this rabble that he subsequently died from his injuries. Several rough scrambles took place and several were injured but, because they were always about forty to three or four, the Nationalists got the worst of it that time.

At the time of the riot I had a contract to dismantle and load a large Lancashire Steam Boiler on to a railway truck. This was the largest boiler I had ever handled. It was eight feet six inches in diameter by twenty eight feet long. I had to roll it about a quarter of a mile to get it to the railway line. I had five Diggers helping me, and am afraid that none of them were unionists.

We were proceeding with the work when we saw about thirty or forty men coming over the hill. They were mostly young. We had heard of their raids on other mines and feared we were in for a rough time. They came up. Their leader, in a blustering manner, wanted to know what we were doing. I explained that I was a contractor and was loading the boiler. They wanted to know whether my men were unionists. I said I believed they were members of the Returned Soldiers Association.

On being informed that it was only a few days' job and, after a short consultation, they decided to allow us to proceed, for which we were very thankful; we did not want any trouble.

A conference was called at the Chamber of Mines which was under the protection of the police. After considerable discussion truce was arranged for about four days, during which the mines were shut down.

In the meantime the police swore-in about 600 Diggers and civilians, brought up a number of protestant police from Perth, and a large supply of rifles, bayonets and ammunition.

One amusing incident occured during this period. A railway official who had loyal tendencies, noticed a case of rifles which was consigned to some unknown man, such as John Smith. He immediately assumed that they were being imported by the rioters and gave the alarm. Of course the police confiscated them. But the fact was that they were police rifles being surreptitiously landed to avoid any scare.

On the last evening of the truce all was ready. 600 fully armed men marched out to the mines at three o'clock in the morning to be housed in the change rooms and other buildings of the mills, all under strict discipline and police control, and warned not to use their fire-arms unless commanded to by their officers.

Because it was a test match between the soldiers and the reactionary section of Labor as to whether the promise of preference to soldiers would be maintained, every man was eager for the fray. A small spark would have been sufficient to raise a big fire and cause many casualties.

The Labor forces assembled at the Boulder Block. They were led by their parliamentary leaders, Mr Collier, Munsie and Lutey, and other union officials.

Collier who was always a shrewd man, paid a visit to the mine to be shown the soldiers and their equipment. He realised what they were up against so went back to his men, told them what he had seen, and advised them to go home for, if they started anything, the soldiers would finish it for them. So very reluctantly they kept quiet.

It was anticipated that they might try some other mischief, but men with bicycles were stationed on all important roads to give warning in case of a rising in any direction. This proved a wise precaution because about midnight, approximately 300 men started to march in from Boulder City to smash up the Soldiers' Institute. Of course the alarm was given before they got half way to Kalgoorlie. Immediately every bell, steam whistle, motor horn, bugle, or other noisy means of raising the alarm was sounded. In a few minutes the army of special constables and soldiers answered the call. The rioters soon discovered that they were again running into trouble. They halted, hesitated and in a few minutes, were again sounding an ignominious retreat.

A small force was constantly kept on each mine in case of a revival of the riot, but nothing happened. The rioters stayed away from work for nearly three months then sneaked back to where jobs were still available for them. Their defeat was effective and crushing. To this day, five years later, it rankles in their minds and, as opportunities arise, they still try to take revenge on the Diggers...

The boiler I had loaded proved to be too high to pass the railway standard. It was too large in diameter to set down between the stanchions on the truck so I had to pack it up with eight inch square timbers, in order to reduce its height. It became necessary to bend the stanchions outward and remove the eight inch packing pieces. This had to be done at the railway workshops, but I had to do it. However, the Loco Foreman remarked.

"When you come down you don't need to bring too many men. I have men to burn down here."

"Right Oh. And what about a few tools and blocks from the repairs wagon?" I asked.

"Oh, yes we can lend you what you want in that line."

So, on going to do the work, I went down on my own. I found the foreman and asked for a couple of men. He asked how many I wanted. I replied that two would do so he called up a boy and said.

"Go down to the running shed and send up some men."

In a few minutes I saw a dozen men, mostly young, coming up. They were skylarking and throwing waste at one another, and their language was lurid and shocking. I remarked that this crowd would be rather in the road because I could only employ two at a time.

"Oh, that's all right." replied the foreman. "You can have a dozen more if you like. I've got men to burn down here."

We got the work done in good time. There were a few good workers among them when they knew what to do. On enquiring I had learned that there were thirty six men in the running shed with nothing to do. The fact was, when the soldiers came back they were reinstated and, for fear of trouble, the men who had taken their places were not dismissed. The railways became enormously overmanned, and I suppose the same was true about all other government projects and departments. One can just imagine what this silly policy cost the long-suffering taxpayers.

Take thirty six men in Kalgoorlie at, say, £4 per week each. This meant £7,448 per year wasted. And because Kalgoorlie is only a small section of the railway department, one can easily imagine that the same thing prevailed all over the state, and that the total cost would be something over £100,000 per year.

How could the railways pay? How could the government reduce taxation with such rotten administration and mismanagement as this? How long this state of affairs prevailed I do not know. All I know is that it prevailed in this disgusting way at that time; one does not wonder that state enterprises are a failure and a drag on the nation.

The go slow policy of the official Labor Party is also founded on this principle. They argue that if one man does only half as much work as he should do, two men will have to be employed to do the job. They do not realise that by a man doing only half his share, he practically doubles the price of the goods, by means of which he limits the market for such goods probably 75%. Thus instead of creating work for another man he is actually throwing another man out of work.

THE WESTERN AUSTRALIAN GOLDFIELDS COURIER.

DEEMING!!

Was an authority on cement, but nevertheless he little anticipated the astounding richness of the Kanowna Cement Leases.

W. A. WYATT,

OF THE

OXFORD SALOON,

Opposite Post Office,

KANOWNA,

Does, however, fully realise the worth of the recent "finds," and he is in a position to offer the public

The Best Glass of Beer in Kanowna

For the moderate sum of **6d.**

NO PIPES USED,

but the life-giving amber liquid drawn straight from the wood.

Don't you forget it.

W. A. WYATT,

Oxford Saloon,

Opposite Post Office,

KANOWNA.

KANOWNA RUSH.

The Leading Butchers.

FOX & HOLMES

ISABELLA STREET,

KANOWNA.

DAILY DELIVERY TO

Camps and Townspeople.

SMALL GOODS A SPECIALITY.

Signs of the Times

A SURE HELP

To the enterprising trader is a well-painted signboard.

GRIEVE BROS.

(OF KANOWNA)

Do not paint the town red, but they will paint you the best shop signs procurable in the district at a moderate figure.

Painting, Paperhanging, Glazing, Graining, Gilding and all Decorative Work

Executed by competent workmen, with rapidity

AND

AT LOWEST RULING PRICES.

GRIEVE BROS.

Opposite Doyle's Hotel,

KANOWNA.

E. B. CASTIEAU & CO.

Livery & Bait Stables,

Opposite Warden's Court,

KANOWNA.

Horses and Vehicles of best description always on hire at low rates.

CARTING UNDERTAKEN.

Horses, Harness and Vehicles Bought, Sold or Exchanged.

Saddlery and Harness of all sorts kept in stock, and repairs of every kind executed on the premises by competent Saddlers.

E. B. Castieau & Co.,

GOLCONDA STREET,

OPPOSITE WARDEN'S COURT,

KANOWNA.

(Associated with W. B. Castieau & Co., Liver Stable Perth and Coolgardie.)

ARMSTRONG'S CYCLE AGENCY.

Head Office: PERTH.

Branches: COOLGARDIE, KALGOORLIE, BOULDER CITY, MENZIES, KANOWNA.

Sole Agents in W.A. for

Rover, Singer, Raleigh, New Rapid New Hudson and Yellowfellow Cycles.

CASH OR TERMS.

FOR HIRE BY THE HOUR, DAY OR WEEK.

Dunlop and B.B. Tyres.

LARGEST STOCK OF BICYCLE ACCESSORIES IN W.A.

Raleigh 1898 Pattern fitted with New Rapid Back Wheel, which has never been known to break.

E. G. & Co.

K.

The above marking may be seen on the hundreds of packgaes of varied merchandise now in stock, and on the road, for the enterprising firm of

Eastwood, Glover & Co.,

KANOWNA.

E. G. & Co. hold the largest stocks in the district, and are in a position to quote

BED-ROCK PRICES FOR

Wines, Spirits,

Beer, Groceries,

Ironmongery, General

Mining Requisites.

The Kanowna Alluvial Fields are the biggest and most genuine thing of the kind in the Colony, and EASTWOOD, GLOVER & Co. are the boss storekeepers of the district, so give them a call for anything you require in their line of business.

Eastwood, Glover & Co.,

(TWO STORES).

ISABELLA STREET,

KANOWNA.

Wholesale Store:—Near DEMOCRAT NEWSPAPER OFFICE.

"Up Town" Depot: — CEMENT STORE, near DOYLE'S KANOWNA HOTEL.

FRUIT, VEGETABLES, DAIRY PRODUCE.

MITCHELL & McCABE,

ISABELLA STREET,

KANOWNA,

Are in a position to supply the Residents of Kanowna and surrounding districts with **Fresh Fruit, Vegetables, and Dairy Produce three times per week** at moderate prices, and of the best quality.

STORE near 'Democrat' Newspaper Office, Main Street,

KANOWNA.

DAILY DELIVERY TO CAMPS AND TOWN RESIDENCES.

GOLD. KLONDYKE Finds are Doubtful,

BUT

Kanowna's Richness.

Is a pleasing Reality.

Residents and Visitors to Kanowna Can obtain the news of the World in the form of the latest Illustrated and general Newspapers, Magazines, &c., &c., at

HUGGINS & FULLERTON'S

NEWS AGENCY,

Larkin-st., Kanowna

The Route to the Post-Office.

All the latest Magazines, Pictorials and Novels obtainable at standard prices.

Stamps, Fancy Goods, &c., &c., Exchanged for Coin of the Realm at O.K. rates.

CHAS. HUGHES,

LARKIN ST., KANOWNA

(Near Armstrong's Cycle Agency).

Produce Dealer.

Hams, Butter, Eggs, Bacon, Honey, and General Dairy Produce.

Lines obtainable in large or small lots at the lowest prices on the fields.

C. HUGHES is a large **cash purchaser** direct from the principal Victorian, South Australian and New Zealand Factories, and has his own Transport Agent in Fremantle, a fact which buyers benefit by, as thereby he is enabled to place the **best** and **freshest goods** before them **at moderate prices.**

DEPOT:—

Larkin-st., Kanowna.

Between Exchange Hotel and Warden's Court.

MANNERS BROTHERS

Mining and CONSULTING ENGINEERS AND DRAUGHTSMEN.

PLANS, SPECIFICATIONS, AND BLUE PRINTS PREPARED.

Mining Plants supervised, erected, and reported upon.

MACHINERY BOUGHT AND SOLD ON COMMISSION.

6 H.P. HORIZONTAL ENGINE FOR SALE.

Agents for Burnell Patent Pneumatic Condensers.

ADDRESS—

BOX 6, KANOWNA.

THE RICHEST ALLUVIAL FIELD IN AUSTRALIA

AT THE PRESENT TIME IS UNDOUBTEDLY THE

Kanowna Cement Rush,

And the

Cheapest and Best Store

ON SAID FIELD FOR

GENERAL GROCERIES & MINING REQUISITES

IS WITHOUT A DOUBT

R. JACKSON'S,

OPPOSITE

DOYLE'S HOTEL.

TEST THE TRUTH OF BOTH THESE ASSERTIONS.

Jackson's Store

is

THE NEAREST TO THE SCENE OF the RECENT RUSH FINDS,

And the Proprietor will be pleased to

Supply the Mining Population.

With all necessaries

At Lowest Ruling Rates for Cash.

R. JACKSON,

Kanowna.

BRANCH STORE AT MULGARRIE.

OPENING

KANOWNA RAILWAY,

KANOWNA.

SEVERAL FULL

BLOCKS OF LAND

FOR SALE,

66ft x 165ft.

SITUATED NEAR

RAILWAY STATION,

And best Central Positions,

The Railway, which will be completed in a few weeks, and the now proved Rich Alluvial Cement adjoining the Town, will considerably enhance the value of property in Kanowna.

For full particulars apply

W. LOWES,

Auctioneer & Sharebroker

KANOWNA.

Kanowna Newspaper - November 27 1897.

SKETCHES OF A FEW WELL-KNOWN IDENTITIES OF THE GOLDFIELDS.

27

Physicians And Surgeons

Just before the boys returned from the war an epidemic of influenza of a particularly severe and deadly type spread through the community.

Unfortunately at this time my eldest son and his wife and family of five children were staying with us, the pigs having got him down on the farm. He was losing his health. This meant eleven of us living in a five-roomed house when the trouble struck us. Nine out of the eleven took the influenza, leaving my daughter and me to nurse the lot. We could not get a nurse or other assistance for love nor money because there was scarcely a house without the dreaded sickness. Many deaths were recorded every day.

I called in on Dr Sparkle, who informed me that he was attending upwards of seventy patients each day. He would come in with a rush, bounce from one room to another, feel their pulses and say "Continue with the medicine", and bounce out again. He was really no help to us because none of the patients was making any progress.

My wife was much worse than the rest and was giving us grave anxiety because she was very low. One morning about 4 a.m. my daughter called me up saying that her mother had fainted and was as cold as ice. She was very much scared and, when I saw the condition of my wife, I practically gave up hope; she was as pale as death, dead cold, and her heart was scarcely beating.

Fortunately the fire was on and the kettle boiling. I took a single blanket, rolled it up into a roll about fifteen inches long then poured a kettle full of boiling water into the interior and kneaded it well through the blanket. I packed her legs in this and it was not long before we saw circulation of the blood increasing. When the blanket cooled I repeated the operation with another and, by the time the heat of this had been absorbed, she had resumed her natural colour, was quite warm, and circulation was fully restored.

She could not have lived another hour in the condition I found her. My prompt action undoubtedly saved her life.

Next morning when the Dr came bouncing in, rubbing his hands together, we told him of the fainting fit. He laughed, saying, "Oh, that's nothing. I saw two people snuff out just like that (snapping his fingers) this morning."

On his journey to the next room he said to me.

"You know, I'm not a physician. These jobs are no good to me. I like operations. Fifty pounds at a time."

This, with his previous remark, so disgusted me that I sent him word next day not to call again.

This episode increased my contempt for the methods adopted by the medical profession, or at least a majority of it. The circumstances were so distressing, and the callousness of the doctor so sordid, that one could not help being disgusted. I was glad to see the last of him and feel that we would be much better left to our own resources. Of course in the case of a death the position may have been awkward, but I was not afraid of this because my opinion was then, as it is now, that the unaided services of a doctor were more likely to result in death than the absence of a doctor altogether. And even under the care of a professional nurse, who slavishly carries out only the instructions of the doctor, the result is too often disastrous.

Seven of the nine who were stricken with this severe complaint had been under the care of the doctor. We had not attempted other means, except in the case of my wife, whose case was really desperate. However, the other two caught the infection, my eldest and youngest sons, aged about thirty, and about twelve.

As soon as they were laid up I gave them doses of Ammoniated Quinine and treated them with a soapy blanket. This consisted of a single blanket rolled into a roly-poly, similar to the one previously described, but filled with boiling soapy water. (There must only be sufficient water to moisten the blanket without making it sloppy). This has the effect of opening the pores of the skin. The soap cleans the pores. After about an hour of this treatment they were sponged all over with warm vinegar and then annointed with warm olive oil. This was the only treatment given to these two. They were up and about in a couple of days while all the rest were in bed from three weeks upwards.

I will use this chapter to describe two other cases in which doctors received surprises. If they did not learn something it was because their methods were so confined and hide-bound that they could not learn from experience, but must be guided by their books and customary practice.

These cases happened at a much earlier period than the influenza cases referred to above, but it was not convenient, for logical reasons to introduce them at their respective periods .

My sister-in-law who was then a young woman about twenty years of age had been playing tennis and, in endeavouring to return a rather difficult shot, strained herself internally. She was brought home in a cab, put to bed, and the doctor called in. Her pain was very severe, in fact, excruciating.

The doctor was, of course, sent for, and was soon in attendance. He examined her carefully. To relieve her suffering gave her a small dose of chloroform and injected morphia. He declared that she had dislocated her ovaries, that it was a fatal accident because she could not possibly live many days under the circumstances. Her only hope was an operation, and even this was a meagre one, but, being the only hope, it would have to be tried. He left a few morphia tablets to be given her one at a time as the effects of the original doses wore off. This would save her from suffering as much as possible. He said he would return at 9 o'clock in the morning, then if she were no better, he would wire to Kalgoorlie (we then lived at Kanowna, twelve miles from Kalgoorlie) for a surgeon. The operation could take place in the afternoon.

Our distress can be easily imagined. Here was a fine, strong young girl with practically no hope of living. In fact, who was evidently doomed to face her Maker within a few days, her only chance the precarious possibility of an operation which would probably cost us £100. What an opportunity for the medical profession. £100 for a few hours' work. To save her we would gladly have spent the £100, but it was poor satisfaction to know that the odds were about 100 to 1 against saving her life.

However, one should never despair. There is an old saying which I have made a sort of motto in my life and that is "Never say die, until a dead horse kicks you". So we got busy. We went to the stable and got a bushel of bran. We started at 8 o'clock that evening as soon as the doctor had gone, applying bran poultices to the lower part of her stomach. We were soon greatly relieved to find that while the poultices were hot her pain was reduced and ultimately relieved. We kept this up for several hours, but found that, as soon as we discontinued, the pain returned. So we resumed our treatment, with the result that it was kept going continuously until the doctor arrived at 9 a.m.

He examined the patient again and asked whether we had given her the tablets. We assured him that we had not. Then he said.

"This is remarkable: her temperature and pulse are both normal, and the inflamation has practically departed. What have you been doing?"

We told him, and he again admitted that it was wonderful, that there would be no need for an operation, but to keep the poultices going. This we did. The poultices were kept going continously for thirty six hours, when it was found the pain dispersed for ever. In a few days the girl was attending to her ordinary domestic duties, and as well as ever.

This treatment cost about two shillings in cash and about thirty six hours loving service gladly rendered. The doctors were robbed of £100, the undertakers of about £50. No wonder the medical profession does not encourage home treatment.

The next case was one of pneumonia. A neighbour of ours was very ill with it and had been under the doctor's care for about three weeks. They were very poor. Their bread-winner being thus prostrated, and the doctor's charges, did not increase their wealth. Their house was small, the ceiling low, and the weather extremely hot, the thermometer registering about 104°F.

The doctor had declared that, owing to the weather and the smallness of the room, the patient could not live more than a few days. He advised that he would have to be removed to the government hospital.

The wife dreaded this very much, feeling that if he went to the hospital he would not have a chance because she knew of two or three who had lately died in the hospital from the same complaint. So she came in and asked us whether we would try to do something for him. We assured her we could not interfere with a patient who was in a doctor's care because if anything unforeseen happened we could be charged with manslaughter. However, she was in such distress, and pleaded so pathetically, that I agreed to come in and see what I could do after the doctor had paid his evening visit, on condition that he was to know nothing of my attempts, no matter what the result.

So it was arranged. The doctor's time was 9 p.m. When he had left I was sent for. On entering I felt the scare of my life. Here was the man in a raging fever, groaning and writhing in intense agony. His temperature was in the vicinity of 106°F and he looked to me as though he would not see the night through.

My first impulse was to decline to do anything, but, under persuasion, I was induced to proceed.

They had a block of ice, and I procured two good soft towels. I wrung these out in the iced water then folding them up to about nine inches square, applied them one at a time to the side where the pain was. These were changed every few minutes. In a short time his groaning ceased and the comfort felt from the cooling effect of the towels was very soothing so that the patient was dozing. After about an hour of this treatment I applied a hot blanket to the feet to infuse life, giving warmth to his circulation, at the same time continuing with the cold towels. After two hours I ceased for a time. His pain showed signs of coming back, so I continued with the treatment (i.e. the cold towels). After another two hours I ceased again, with the same result, so I continued for another two hours and, at four o'clock in the morning, dried him and rubbed him with warm olive oil, changed his bed, then left him sound asleep. His sleep lasted until the doctor came at 9 a.m. This was the first sleep he had had for days.

The doctor examined him. The pain had gone, inflamation had vanished, his temperature was normal at 98.2°F, his pulse was also normal and regular. The doctor was amazed and declared he could not understand it, that a marvellous change had taken place during the night, and he would not need to go to hospital now.

Of course he was very weak for some time after so severe a shaking, but in less than a week from that date he was on the front verandah smoking his pipe - the doctor lost another patient. No wonder they do not encourage these methods, for, if these simple remedies of nature and common sense were applied they would have less patients and less revenue.

Before I close this chapter and leave for good and all this interesting topic of the medical faculty, I will just mention a few more cases of simple and effective remedies. For the purpose of this story is not so much to provide an attractive story as to hand down for posterity to those who care to read these lines, hints and experiences of my own life which may be useful in saving life and money as well as pain and suffering for perhaps many who would otherwise be the victims of what is either ignorance or deliberate imposition in medical practice.

While doing some work one cold day I got my clothes wet and, when changing them before a fire, hung my wet pants on the mantlepiece, keeping them in position with a 200 ounce gold scale weight. I inadvertently touched the pants, pulling them off the shelf. The weight fell fair upon my big toe.

The wound was exceedingly painful and gave promise of a week, at least, without being able to put on my boot. The bone was not broken, but the nail was blackened and the toe severely bruised. Professionalism would have bathed the toe for five minutes or so in luke warm water then painted it with iodine, wrapped it up, and advised not using it for a week at least. A week's work would have been lost and a week's pain suffered.

What I did was to get a basin of very hot water and insert the toe in it. It was hard to bear at first, but I reduced the temperature until it was bearable, then gradually increased it as the pain eased. I had a kettle of water on the fire. I kept the toe in the water until the pain was all gone, never to return. I did not notice the time this bathing lasted, but it was probably two hours. Even if it had been three it was worth while because from that time I had no more pain or inconvenience. I put my boot on in the morning and went to work as though nothing had happened.

I would have you note that in this, as in others, the treatment was persistent and continuous until the pain ceased, never to return.

Another case of my own experience was with a steam boiler in which the water gauge glass broke and steam and water was escaping from the boiler in a cascade. The fireman had run away from it, but it had to be stopped in order to keep the plant running. I threw my coat over my head and rushed in and shut off the cocks, thus saving the situation. I, however, got my hand badly scalded.

If I had called the doctor he would have put some oil on this, bound it up in cotton wool, and dressed it every day for some weeks, during which time I would have suffered untold pain and inconvenience.

I didn't call the doctor. I simply immersed my hand in a bucket of the coldest water I could obtain. As soon as I put it in the cold water the pain ceased, so I kept it there until the pain ceased, to return no more. I think it took about an hour; persistence again. However, I was able to use the hand immediately after, and suffered no further inconvenience.

On one occasion one of the family got a terrific pain all over the head and face. It was undoubtedly Neuralgia. We tried several things, such as a bran poultice at the back of the head, cold cloths on the forehead, linaments rubbed on, but all to no purpose; the pain persisted. I then took a small blanket, made a roly-poly as previously described, kneaded the hot water through it, then wrapped the whole head in it, leaving only the nose out for breathing purposes. In less than an hour the pain had gone and there was no return of it.

What joy it is to be able to relieve one from suffering in so complete and simple a manner. And what a pest it is to have to stand back and see folk suffering and dying all around one, almost every day, and be unable to help because it would be a criminal act to interfere where a medical man and a subservient professional nurse are pottering around with injurious physics and useless Antiphl.

I will wind up this chapter by referring to a few simple remedies which, if applied in time, will nearly always save further complications and long illnesses.

As soon as you feel an aching in all the bones of your body you know you have influenza coming on. Take a teaspoonful of Ammoniated Quinine and pour water into it until all the precipitate is dissolved and the water quite clear, then drink it. Repeat the dose in about two hours. Two or three doses will prevent the attack coming on. If, however, you have left it too late i.e. two or three days, this treatment will not be sufficient. Then you must resort to the soapy blanket previously described. Ammoniated Quinine gives some folk a headache. A couple of asprin tablets (5 grs. each) will invariably remedy this.

At the first symptoms of a sore throat, which would ultimately result in Quinsy or a similar complaint which is a very painful and distressing illness, take two teaspoonfuls of Peroxide of Hydrogen and mix it in an eight ounce bottle of water. Use this as a gargle every hour or so for three or four hours, and that will be the end of the sore throat.

Again, on feeling the first symptoms of bronchitis, which are a pain in the chest and a rasping cough, take a small quantity of mustard and mix it with olive oil into a very thin paste, seeing that all the mustard is saturated, and rub this well into the chest, say at night and again in the morning. You will probably have no more trouble from this complaint. If, however, you have left it too late, this will not fully cure the complaint; then resort must be made to the bran poultice and cold towels referred to in chapter twelve.

When I had the fever in Broken Hill the doctor was dosing me with Quinine, presumably for the purpose of reducing my temperature. I remember on one occasion my temperature had been at 103°F and the next visit it had dropped to 102.8°F. The doctor was quite pleased with the "wonderful improvement" and remarked that he did not wish to reduce the temperature too rapidly because it was dangerous. We really believed him then and, of course, were pleased with the result. But, when next day the temperature was back to 103°F again, our hopes were dispelled.

The fact was that the doctor did not know how to reduce the temperature, and his quinine was quite futile as a remedy.

My experience has been that the sooner the temperature is reduced the sooner will the patient get well. I find that this temperature can be reduced to normal in a few hours by getting the right application, which mostly consists of getting the skin to act by its pores opening. Evaporation of perspiration will do the rest.

Brother Trethowen.– Praise the Loard, Brother, eddn't she a beauty?
Brother Grub.– Eddn't she? Brother, God put un theere, we find un; as soon as I seed her, I knoo oo she belonged to.

28

Gold Stealing

THE MINE DETECTIVE

The jewellers and fancy goods dealers were doing quite a good trade and the barmaids, waitresses and other attractive morsels of humanity were richly endowed with various articles of adornment in consequence.

Tim O'Shannessy, a typical son of the old sod was waiting in the Jeweller's shop while the assistant was fitting a glass in his old English lever silver watch. Tim always had his eyes and ears open for stray bits of information which might help him in the execution of his duty. He was not a publican or a policeman as his nationality would imply, but his calling was so nearly allied to the latter profession as to be not inconsistent with the traditions of his race. He was a mine detective. This was an occupation rendered necessary by the well known prevalency of gold stealing and the resultant illicit dealing in that precious metal. For a miner, or for that matter a mine operator of any class, does not consider it a crime against his conscience or moral code to appropriate to his private use any stray pieces of gold or amalgam that come within his reach, although the Majesty of the law is so unkind and unreasonable as to consider such acts, felony of a high degree. So much so, that if a person is found in possession of gold or gold bearing ore he must prove that he came by it honestly or the obstinate Law will consider him guilty.

Mining Directors also are reputed to meanly begrudge the reduction of their dividends by this nefarious trade as they term it while a Cornishman asserts, "God put'n there and him as finds 'n owns 'n".

So Tim O'Shannessy found a job to his liking, for he was not fond of hard work and it was his duty to obtain information of such cases and should further proof of guilt be necessary, by application to the underground manager, he could find employment in any capacity (irrespective by his skill or lack of skill) in any part of the mine.

Tim was in luck today for while waiting for his watch he observed a lady examining some silver backed brushes, hand mirrors and such goods, when a gentleman friend of hers entered and Tim heard the following conversation

"Good afternoon Mrs. Leaming. Lovely day, What! giving the Doctor a present." "Oh, no," replied the lady. "I am merely satisfying my curiosity, my maid's sweetheart has just given her a present of similar articles to these and I have just ascertained that they cost him seven guineas.

"Lucky little girl," retorted the gent. "A very generous lover, eh? He must be in rather a good situation". "Not at all." said Mrs. Leaming, with a meaning smile, "He is only a miner but he works in a very good mine."

Tim smiled inwardly and departed with his renovated watch, "This is a fair cop this time, I am right on to distinguish myself," thought Tim as he walked along.

He knew Dr. Leaming's home and decided to watch it for Mary Ann's night out. He did not have long to wait for a couple of evenings later he noticed a man hanging about the neighbourhood until at about eight o'clock the maid joined him. They went to the picture show in the next street so Tim had a look at the show too, which filled in his time much more pleasantly than proved the subsequent hour or so required by all true lovers to say goodnight.

However the time wore on and eventually resulted in Tim following his quarry to his camp. There was no doubt he was on day shift as either afternoon or night shifts would have preluded his visit and his subsequent prolonged farewell greetings, so there was no more to do but be in the vicinity early in the morning in miner's attire. Tim was there betimes and followed the innocent to the mine, where his name and position were quickly ascertained from the manager. He proved to be Fred Trethowan a Cornishman about 30 years of age. He had not been many years in the State and had lost little of his native manner, he wore a goatee beard and in conversation had that expression in his grey eyes, typical of Cornishmen, which signifies a consciousness of unlimited inherent wisdom as much as to say "I do knaw, I do and if I don't knaw no one knaw."

The next morning Tim was sent to join him as his mate, his former assistant being sent to another portion of the mine.

They worked together for nearly three weeks without any sign of pilfering in fact very little gold had been visible in the face. They were very sociable as mates and talked on all sorts of subjects especially on socialism and politics, while Fred often tried to draw Tim out as to his willingness to help himself while Tim had to act the part of honesty personified; for we must give him the credit of being man enough to avoid tempting his intended victim into crime. Fred would say "We do see some nice bits of gold here at times, we do. There be some as 'ud think they owned 'un, and there be some as say they be fools as don't think so. What do 'ee say, mate? "

Tim's answer would be cautious for instance "I'm paid to work for the Company and I loikes to do me dooty." Fred warned by Tim's reticence had evidently decided not to take him into his confidence for one afternoon Fred's pick point struck something tough and spongey and experience warned him it was a lump of gold although covered with earth. Fred had thought out his plan for such a situation. When 'firing out' that is blasting in the face before lunch they always retire to the plat about a hundred yards distant to avoid the dynamite fumes while eating their "crib" so Fred said "Say Tim, blawed if I didn't leave our detonators down at the plat at crib time, skip along and get 'em, will ee? or someone else will nab un." Tim was not to be fooled so easily but he proceeded on his journey until he reached a bend in the drive where he blew out his light and returned to watch proceedings from a few yards distant in the dark.

Fred's pick was going harder than it had previously moved when duty only was the impelling force and in a few moments a beautiful slug of gold was unearthed. Fred's face was beaming and Tim heard him soliloquizing thus "My word you're a beauty, ye are so soon as I saw 'ee I knawed who 'ee belonged to, yer mine my dear, yer mine." He carefully wrapped it in a piece of bag and buried it a few inches deep beneath the ore in a truck and with a piece of chalk made a small cross on the truck door. Then Tim went round the bend lit his candle and returned and growled at Fred for sending him on a fools errand as he could not find the detonators. Fred admitted he might have made a mistake adding "Maybe they be in yonder box lad".

The braceman was a party to the scheme and seeing the mark when the truck came to the surface, and knowing its meaning, unearthed the parcel and hid it under the tramway on the mullock dump.

At knock off time all went through the change room where each miner has to strip off his working clothes in one room walk along a passage in nature's garb and after a bath, don his outdoor clothes in a second room. His clothes and tea-can are diligently searched by the officials but no gold is ever found in them. Tim's work now began, he never lost sight of Fred from the time they left the change room for he had to catch him with the gold in his possession in order to obtain a conviction.

About 10 o'clock that night Fred and the braceman made for the mine and very cautiously avoiding all watchmen and strangers the braceman climbed up the dump while Fred kept watch. A few minutes later they were together again and finding all clear started for home when suddenly they were confronted with the sight of a revolver pointed at them and the words "Hands up". They were startled almost out of their senses, but as soon as these valuable faculties returned Fred recognised Tim, and suspecting the truth had to resort to subterfuge "Hello Tim old pal," said he, "What's the joke, are you playing bushranger? we ain't got nothin' nohow." "No" said Tim, "I'm a private detective, I am, and I saw you plant that little slug this afternoon, and I've followed yees ever since, and now I've got yees both, so come along wid me to the manager and no shenannigans."

"How much do the company pay ye for this here job?" said Fred with the cunning leer in his eyes. "Five quid a week," answered Tim, "And I'm going to do my dooty by 'em."

"Well you are a chump, matie, to do this here dirty work for a tucker wage, when a fortune be staring 'ee in the face, look here Tim, we've got about two thousand quid in this here little parcel and you're on a third share, fancy five quid a week for - let me see - nearly three years. Get along with 'ee, 'ee be' half mad be 'ee."

If it had been a paltry ounce or two it would have been bad for Fred and his partner but his was beyond human nature's powers of resistance. Tim had seen the nugget in the afternoon and guessed its weight at 600 ozs. So merely said, "Get off the lease as quick as ye came ye divils," and disappeared.

Fred married Mary Ann a few months later and put her up in a nicely furnished house as Mrs. Trethowan, and Tim after working as private detective for three years retired and bought a hotel in Sydney for twenty five thousand pounds; and paid cash.

Editor's Note: Whilst the stories in this chapter are true stories, the names of the individuals have been changed in respect for their descendants

THE COMMISSION

It was the startling headlines that did it, everyone knew that illicit gold dealing was rife in every quarter. Both the stealing and buying were going on with impunity, but one morning, the mining magnates, the police and the legislators were startled when on opening their morning paper they saw the large black headlines.

AMAZING FIGURES
A MILLION OUNCES OF GOLD STOLEN
EVERY YEAR

An itinerant newspaper correspondent had made the statement that the mines were being robbed of this enormous amount of gold every year and his statement was published under this headline. He also drew attention to the inadequacy of the law and the inefficiency and absence of authority of the police to deal with the question.

The effect was like a match to a powder magazine. Iky Mo in London bewailed their loss and the cable lines were overwhelmed with messages of instructions to their managers.

The police departmental honor was at stake and the government awoke from its wonted somnolence and immediately inaugurated a Commission to enquire into the scandal. It required a week or two to appoint the Commissioners and another week or two for them to form their plan of campaign. By the time they managed to summon their witnesses, of whom the police had a long list, there had been a large exodus of assayers and metallurgists whose shops were deserted and no one seemed to know where they had gone.

A few who, everybody knew, were illicit gold buyers remained but these were mine owners with rich ore in their mines or they had residue dumps and cyanide plants.

So the Commission after sitting for three months and filling 7000 pages of foolscap with evidence had not sufficient information to bring a case against even one individual and the only result was the passing of a law making it incumbent on any individual, found with gold or gold bearing ore in their possession, to prove that they came by it honestly or the law would consider them guilty of felony. Of course many mines' employees were very nervous during the lengthy deliberations but were soon reassured when the ineffectual efforts of the Commission became apparent.

Joe Portman owned a mine which always showed free gold in the face and it was astonishing how rich and consistent his monthly returns were. There was a slight decline in his returns during the first week of the Commission but his clients where soon reassured.

Fred Trethowan was afraid for a while that his occupation was gone but he called on Joe Portman and started the conversation with "Good mornin' Joe how do this here Commission suit your business?"

"Tip top" said Joe" Just the thing. It has driven all these little tin pot dealers to Jericho and now I see my trade will be doubled."

"Then" enquired Fred "If I happen to have a little nest egg in my pocket I can leave it here eh?" "Fetch along a bolly ton of it" replied Joe. "Biant 'ee aheard the cops will lay a trap for 'ee my lad" was Fred's next query "No fear" rejoined Joe" My mine is ready for them and most of my customers are good men. I have just put through 800 oz for one of the mine managers, see."

So Joe's business prospered and the police were powerless to catch him as Joe was much more cunning than they thought.

One afternoon Joe and his trusty assistant were in their office when a man came in with a bag of specimens of sulphide ore of a peculiar class which was common to only one mine in the district. The man was quite a stranger to Joe and was rather a disreputable looking character when he produced the ore and asked for a sovereign for it. Joe's suspicions were aroused and he asked his doubtful customer where he got it? Getting no satisfactory answer he asked him who sent him with it? The man began to fidget and displayed suspicious measures which roused Joe's ire. So he and his assistant thrashed him till he was quite helpless, stowed the ore in his shirt and carried him out and threw him in the back lane.

A quarter of an hour later three policemen who had been vainly awaiting their accomplice's return rushed in with a search warrant and were cheerfully received by the smiling Joe and invited to search to their heart's content. And they did search high and low and in every nook and corner but found nothing of an incriminating character and left with rather crestfallen airs.

They subsequently found their decoy in the lane and conveyed him to the hospital for repairs while Joe continued to smile and carried on his business for some years after. He is now touring the world as an Australian mining millionaire.

CAUGHT

Tim O'Shannessy the mine detective had not yet obtained any convictions and his reputation as well as his job were assuming rather precarious conditions. Mac the mine manager had already abused him and accused him of being a fool, and so Tim had to execute himself in the direction of laying some of the gold stealers of no importance by the heels.

True, he had discovered several of the big gangs, but these had always made it worth his while to turn blind and deaf, with the result that Tim's income had become much greater than double the biggest Mine Manager's salary in the district but unless he made his position good by obtaining a conviction he would lose his billet and in consequence his connection with the gold stealing gangs and the profits.

So Tim set his energies towards catching one of the lesser lights. One evening as he was watching the men passing from the pit's mouth to the change room he noticed something like a piece of ore drop off a man's boot on to the ground and when all had passed into the change house he walked over and examined the stone which proved to be a piece of ore showing free gold so he replaced it where it fell and ensconced himself behind an ore wagon in the vicinity and watched that store for several hours until after dark a man came along and picked it up and put it in his pocket.

Tim was beside him in an instant and showing his authority took him to the change room where he was searched and the piece of ore found on him. He was charged with being in unlawful possession of gold bearing ore and asked what he had to say for himself.

He replied that in passing across the property he saw this piece of ore, picked it up and was intending to take it to the mine office when O'Shannessy accosted him. O'Shannessy then said that he had seen the ore drop from the leg of the accused's trousers in the evening and had watched for him coming back. His working clothes were then examined and it was observed that the bottom of the trouser pocket was cut out, evidently for the purpose of dropping the ore surreptitiously to the ground. The police were then telephoned and he was "given in charge".

His name was Tom Wallis, a man of middle age, rather poorly clad and of bilious appearance and looked half fed. He was very depressed and begged to be allowed to go home as his wife and children would be unprotected. His pleading was in vain and he was locked up and a search warrant obtained for his premises. Two policemen were dispatched to make the search.

The police found his home to be a poorly furnished five-roomed house, his wife a shabby woman pinched and haggard with a dragged down appearance and 8 children of all ages from 16 downwards to a puny baby in her arms. She had been a strong good-looking woman in her brighter days but poverty and hard work had made her prematurely old. When the police informed her of the circumstances occasioning their visit, she broke down entirely, became hysterical and the children clung all round her crying and wailing in a pitiful way and the poor awkward bobbies did not know what to do to help her. They however laid her on the bed and proceeded with their search so thoroughly that they found several more pieces of ore in an old outhouse together with a mercury bottle, dolly pot, an old rock drill which had been used for a pestle, and a rusty prospector's tin pan. These they confiscated and carried off to the police station.

Wallis was brought before a couple of J.P.'s in the morning and evidence was taken on lines similar to that in the change room and the prisoner was persuaded to take his trial at the next sitting of the criminal court, bail being allowed. Two of Tom's mine mates managed to find bail and he was released.

The court would not sit for two months and as Tom was taboo as a miner in the whole district, the poor family had a desperate struggle for existence during that period. The eldest boy was earning a few shillings a week and two others sold papers while Tom borrowed a "shaker" and went out prospecting for alluvial gold but found only a few pennyweights.

So the time wore on to the date of the trial, but it was a sad and miserable time often resulting in the parents going to bed without supper so that the children might have what little there was to eat. There was nothing much in the house to sell and the rent had to be paid to keep the poor roof over their heads. Some kind friends had subscribed a few pounds and had hired a cheap lawyer to defend the miner while the mining company was prosecuting the case with vicious vigour and had secured the best counsel available. A firm of assayers had also been engaged to value the stolen ore and experts to prove that it was similar to, and had come from, the mine where Wallis was working.

So everything was prepared for the day of trial. The elder five of the children were sent off to school or work and Tom, his wife and the three youngest trudged along to the court to answer to his bail with heavy hearts and failing hopes for this was like no other felony, in which the law had to prove a person guilty. Poor Tom was already adjudged guilty according to the new law, had to prove his innocence and everyone knew beforehand it was a hopeless task.

The jury was empanelled and consisted of twelve tradespeople who tried to look stern and wise. The judge looked sternly happy as he gleamed at the prisoner in the dock who was pale and haggard and elicited the sympathy of the miscellaneous spectators who find a pleasure in attending law courts and observing other peoples' misery. The Crown prosecutor in opening the case described it a most heinous offence and one that it was the desire of the law to "stamp out with a firm hand". They would call witnesses to prove that the prisoner obtained the ore from the mine where he was employed, and that it was of very great value, that the prisoner had been previously stealing gold ore as its presence and that of appliances for extracting the gold would show.

Tim O'Shannessy was the first witness and he swore that Wallis was the man who dropped ore and subsequently picked it up and described his method of arresting the prisoner. The next witness was Mr Colville de Snook the assayer who had assayed the ore and estimated its value as 37 oz 2dwts 3 grs per ton. In reply to the Crown prosecutor's question Mr Snook admitted that this was very valuable ore. Mr Strut, the lawyer for the defence, then asked the weight of the ore found in the prisoner's possession and this was given as 10 lbs. "Good" said Mr Strut "Now you say this is very valuable ore, what would be the actual value of this small parcel?" "About 14 shillings." "Good, now by these crude methods of extraction, what would he be able to recover?" "Probably a little more than half." "Good say seven or eight shillings." "Yes." "Good, thank you. You may sit down." The Manager and underground Manager gave evidence to show that the ore was similar to that in the mine and dissimilar to that in other mines.

Wallis went in the box and stated that he did not steal the ore, that he chanced to be passing across the property and saw the stone and thinking it of no value picked it up. The other stones he found in different places and as his wages were quite insufficient to keep his family he often did prospecting to try and earn a few more shillings to help to supply food for his large family. He pleaded for mercy on the grounds of his previous good character and the fact that he had a large family who would starve if deprived of his support. This was a fine opportunity for the judge to work off that old platitude "You should have thought of that before" and he smiled at his own wit and wisdom.

Mr Strut put Mrs Wallis in the box for effect and the poor broken woman could hardly speak for weeping. She said her husband had always been good and kind and patient and she was sure he would not steal. He had worked hard to support his family but with broken time and ill health they were often without food before pay day came round again. She tried to proceed, but broke down entirely and would have dropped her child from her arms but for the kind intervention of the policeman who supported her out of the witness box.

Then followed a strong speech against the prisoner, a weak speech in his defence and a summing up dead against him as a criminal of the worst kind. The jury after 40 minutes retirement returned a verdict of guilty, with a strong recommendation for mercy owing to the pitiful circumstances. This the judge ignored and sentenced him to six months hard labor.

Wallis was dumbfounded for an instant then fell forward with his face on his arms on the dock and groaned with agony as he thought of his family's future while the screams of the wife together with the wails of her clinging children made the customary "silence in the court" sound like a sacrilege while the police removed the prisoner and the other causes of disturbance.

So is justice dealt out by the lordly magnates to the humble poor. Oh, you poor miserable sinners, who pilfer paltry pence, in desperation when poverty and hunger pulls your heart strings and turns you with the wailing of your babes.

Thus are you hounded down while those who persecute and judge you deal in wholesale robbery and swindling schemes, robbing rich and poor alike. They flout every law and whilst rolling in magnificent luxuries, ignore the cries of the struggling toilers which faintly reach them in their canopies of power and splendour.

MAC

McClusky was a mine manager and according to his own estimation a man of some importance. How he ever became elevated to the rank of mine manager of a big mine was one of those problems that mining scientists and students who had spent large sums of money and much application on their education and had subservient positions to him, tried hard to solve but gave it up. It was his personality that saved him from remaining a common laborer. He was a bully, a braggart and an egotist and displayed his lamentable ignorance when in conversation with, or browbeating, the unfortunate scientists abovementioned, but mining directors being, as a rule, quite as ignorant of mining science as McClusky, mistook his bluff and brag for wisdom and being partial to slave drivers took him to their bosom.

In order to humble his metallurgist, McClusky would say in a deep bass voice "Your blank blank residues have gone up hagain,' ow hoften 'ave hi told you, you must get more blank air hinto your cyanide solution, you hought to know that the oxygen of the atmosphere is wanted to dissolve the gold."

The metallurgist would be so amazed at this libel on mining science that his efforts to suppress his mirth deprived him of a ready reply and McClusky would strut off with his chest out and his thumbs in his arm holes proudly conscious of having made a hit that time. He was a stickler for discipline so much so that he would never listen to reason or allow any "back talk" as he called it. This quality frequently led him into trouble which no one minded so long as McClusky was the victim but infortunately this was not always so. No number of lessons however seemed to teach McClusky that he was not the embodiment of all wisdom.

On one occasion operations ceased when the ore conveyor which removed the broken ore from the rock breakers at the pits mouth to the grinding mills, was out of order, and was being repaired by a mechanic.

McClusky was underground and saw that hauling operations were delayed and the plat was full of ore wagons ready to go to the surface. He immediately jumped in the cage and signalled to be hauled up and on arrival at the crusher station he bullied the man in charge, cursed him for a fool and then ordered him to start the machine. The man endeavoured to explain the position but McClusky as usual brooked no argument and repeated his order in a louder voice and with more force and profanity. Still the man hesitated when McClusky shoved him roughly aside and pulled the lever which set the machinery in motion. He had hardly time to stand back and assume his wonted attitude of importance when a yell was heard in the distance. When all hands rushed to ascertain the cause they saw the mechanic, who had been fixing a pulley on the belt conveyor, lying on the platform in a faint, his right arm crushed to a pulp and the terminal of the conveyor smashed up and covered with blood.

McClusky was a very quiet man for a day or two afterwards but was greatly relieved when the enquiry committee classed it as an accident.

Our hero was never happier than when he was in a bar parlor with a crowd of professional men who tolerated him, liberality with a man of this class was compensation for his lack of breeding and manners and the bravado of him "shouting" illustrated his generous salary.

On one of these occasions little Dr Leaming together with several other gentlemen and McClusky (not included in the latter category) were spending a convivial evening. The Doctor's prize collie dog was of the company although not indulging in the festivities but was content to receive the petting of the various gentlemen and mistaking McClusky for one, sought his approval. But McClusky to show his importance kicked the dog on the ribs uttering a gruff "Ger out of that, you brute". The Dr, who was only half McClusky's weight but had been to college, was at him in a flash and slapping him on the cheek with his open hand said, "How dare you kick my dog? Defend yourself." McClusky was on his feet in an instant and a magnificent exhibition of brute force opposed to skill and training interested the spectators for one round only, at the expiration of which McClusky had a black eye and several facial bruises and cuts and lay at the feet of the doctor; a very humble spectacle indeed.

The most amusing part of this episode was next day McClusky attributing his misfortune to a buggy accident when the true story had had twelve hours start on the fable.

There was only one man on the mine who could manage McClusky and he was the underground foreman. He was a yankee named Silas Wren. He could swear and bully just as well as McClusky but had the redeeming qualities of a keen sense of humor. He understood human nature and got the best out of his men by humoring their strength and their weaknesses so when McClusky was on business underground or elsewhere with Wren, he was, as near as his nature would permit, civil, for if he bounced he received bounce and if he was uncivil he was met with sarcasm or a joke as the spirit moved Wren.

There had been a good deal of talk about gold stealing and McClusky had failed to trace any of it although he often told stories of how he caught the thieves and how he handled them. He had got it into his head that it would help him if he "sacked" all old hands and being below on a visit of inspection with Wren he observed some men who had worked in the mine stopes for a long time. He said to Wren. "'Ow long has those chaps been 'ere Wren?" "Wall", replied the foreman, "They've been here quite a while. Guess they've been here right along from the start". "Don't you think they've been 'ere too long?" was his next question "You'd better get rid on 'em 'adn't you?" "Wall I guess naat." replied Wren, "These men have all gaat comfortable homes, they're gaat wives and families. They're gaat pianers, gold watches and chains and good furniture. Their wives have goat gold watches and chains and laats of jewelry. They've gaat gramaphone, silverware for their tables, good bank accounts and all sorts of comforts and they are satisfied. Naow if we quit these parties and put on others I guess they will want all these things. I just guess we'd better leave them right here where they aire". McClusky for once was convinced by argument and I guess he left them there.

BETTING

Horse racing, whippet racing and other similar sports were so popular and betting was such a source of revenue for the large number of "sports" who were swarming the town that the tradespeople owing to bad and doubtful debts had reached the limit of the banking companies forbearance in the matter of overdrafts and had induced the government to pass a law making betting in the shops and in streets illegal.

This did not minimise the business much, but the bookies ceased giving cards and clients were required to leave their names with their money and trust to the bookies honor for the dividend should their judgement prove correct.

The police were therefore compelled to obtain other evidence in order to secure a conviction, with this object in view Sergeant McManus invoked the services of a local hotel loafer to go into an alleged betting shop and make a bet for which purpose the Sergeant gave him two half crowns two florins and a shilling all of which were marked. P.C. O'Dowd was a witness to these marks.

And the beer eater entered the shop presumably made the bet and came out.

And immediately after McManus and O'Dowd raided the shops, found the marked silver, and subsequently summoned the bookie to appear before the court.

When the case came on the Sergeant conducted the prosecution and the bookie engaged Mr Hansen a rather clever lawyer to defend.

The decoy was put in the witness box and in answer to the Sergeant's questions told his story of how he received the silver from the police and made the bet with the defendant. Mr Hansen asked a few unimportant questions and the witness stood down.

P.C. O'Dowd then gave evidence as to the identity of the silver which Mr Hansen did not dispute.

Then the bookie was put in the box and on being asked by Mr Hansen if he remembered the last witness coming into his shop on a certain day answered. "Yes What happened? He came up to the counter and asked if I could oblige him with half a sovereign in exchange for ten shillings in silver and I gave it to him." It was one mans word against another so the magistrate had no option but to acquit the defendant.

When the Sergeant met the lawyer after the court he said, "My oath Hansen you are a beautiful liar." "Well" retorted Hansen, "If I were a Sergeant I would not have been such a fool as not to search my pimp when he came out of the shop and allow him to beat me for ten bob!"

Cr. H. Raven.
Cr. E. Harris.
Cr. W. Edwards.
Cr. W. G. Manners.
Cr. H. L. Barnard.
KALGOORLIE · MUNICIPALITY
Cr. A. C. Wall.
Cr. E. P. Broadhurst.
Cr. F. W. Allsop.
W. R. Burton
Mayor
1919 ~ 1922
C. E. Eccles, J.P.
Cr. H. Seddon.
Cr. W. Schwan.
Cr. W. Morris.
Cr. E. E. Brimage.
Cr. S. Madorsky.
Cr. H. W. Davidson, J.P.
T. F. Mackay photo
The MAYOR, COUNCILLORS & TOWN CLERK

THE

Kalgoorlie Miner

PUBLISHED DAILY.

MONDAY, MARCH 7. 1921.

ITEMS OF NEWS

No Licence Campaign.—At last week's meeting of the committee, presided over by Mrs. J. Reid, it was reported that ratepayers' meetings had been arranged to be held at the Boulder Town Hall and the Halfway Hall. Crs. W. G. Manners and H. Seddon were elected as vice-presidents of the movement. The treasurer reported on the collection at the Kalgoorlie Town Hall meeting, the proceeds handed over from the Prohibition League social, and a further lump sum donated by the Prohibition League. Further arrangements were made for street meetings.

Cr. W. G. Manners, the newest acquisition to the Kalgoorlie Municipal Council table, evidenced at the smoke social to Mr. Charles D. McKenzie on Saturday night that he has a gift for characterisation in freehand drawing that should make him a welcome guest at similar functions. Armed with a piece of chalk, and using a blackboard for his operating table, he dissected the facial characteristics of Jews, Gentiles and Confucians with much skill and deft touches of humour. His running fire of comment added greatly to the enjoyment of his exhibition. The test of character by handwriting was applied in a novel way by Cr. Manners, who turned such words as "Cohen," "Irish," "Jap," and "Scot" to hilarious account by working the writing into the counterfeit presentments of types of the particular nationalities Using maps in illustration of his methods he scored an immense success by transforming the map of Australia into a sketch of a typical "Digger."

THE

Kalgoorlie Miner

PUBLISHED DAILY.

WEDNESDAY, FEB. 25, 1920.

THE IRISH QUESTION

Perth, Feb. 24.

Archbishop Clune was the principal speaker at a meeting held in the Hibernian Hall last night to inaugurate a campaign in this State in furtherance of the cause of self-determination for Ireland. Mr. Walter Dwyer presided over a crowded attendance, and the speeches, which were made by Archbishop Clune, Messrs. Collier and Walker, M's.L.A., and others, were eloquent and impassioned. The following resolutions were carried unanimously:—(1) "That we Irish Australian citizens and sympathisers with Ireland's cause affirm the right of the people of Ireland to choose their own form of government and to govern their country without interference from any other nation. We endorse Ireland's appeal to the nations for international recognition, and we pledge our support to Ireland's chosen leader, Eamonne de Valera." (2) "That we publicly condemn the operation of martial law upon the people of Ireland, whose only crime consists in the determination to be free; and further solemnly protest against the ruthless campaign of slander and misrepresentations shamelessly organised by the enemies of Ireland against the Irish race at home and abroad."

It was decided that a West Australian fund be opened. The chairman remarked that there was already in hand about £215.

29

"I Hate England"

The Great War taught us many lessons and, amongst others, taught us that a very large proportion of the Irish people were not only disloyal to England and all that was English, but were a great menace to the nation. The fact that they were a portion of the nation, fighting against the nation, constituted a greater danger to us than a much stronger enemy fighting us face to face.

It is easy enough to understand the German hating and despising Englishmen and all things English because they had been taught in their militaristic training that England was their great rival, their most powerful enemy, and it was their business in their aim for world dominance to smash England or be smashed themselves.

But here was a people living on English territory, for whom England had done almost everything possible; had spent millions of pounds in her endeavour to pacify and make a contented people, all to no avail. The more England did for Ireland the more antagonistic Ireland seemed to become. So wide-spread was this hatred, so universal, that Irishmen living under the free institutions and happy conditions of the British Dominions were imbued with the same discontent, the same disloyalty, and the same unreasoning hatred which enabled them to accept British bounties with one hand while endeavouring to stab Britain in the back with the other.

The position seems so anomalous and untenable that one is influenced to seek the cause of it. There is surely something wrong in the Irish training or education which engenders this inconsistent attitude. I know many Irish people. Even many of those who manifest disloyalty are kind, good hearted people, generous to a fault, brave and hardworking. They seem to possess the fundamental principles of quiet, decent, gentlemen until some incident occurs to stir up that spirit of disloyalty. Then they seem to lose all sense of proportion and become the raging maniacs prepared to hit every British head that comes within range.

My object in writing this chapter is to point out what I think is the cause of this kink in the Irish character, and to suggest a remedy which I feel would, if carried out for even one generation, revolutionise the Irish character and unite these two great people in a harmony which would not only tend to make a great and lovable people happy, but would so strengthen the Empire whose life would be prolonged indefinitely and its greatness and glory enhanced beyond measure.

My first inkling of this idea was conveyed to me at an election held some years ago in Kalgoorlie. The member for Kalgoorlie, Mr P. Mahon (an Irishman), had been expelled from the Federal Parliament for disloyal utterances. He offered himself as a candidate for re-election. A man named Foley (a Wesleyan), was opposing him. I had offered my services in the interest of the latter and was scrutineering at the Railway Institute.

At an adjacent table was a typical Irish woman working in the interests of Mahon. She had her son with her, a small lad about twelve years of age.

P.O. BOX 33
TEL. 121

W. G. Manners

CONTRACTOR, MINING AND MECHANICAL ENGINEER,
PATENT AGENT.

PATENTS,
TRADE MARKS,
COPYRIGHTS,
Applied for in any Country.

PLANS,
SPECIFICATIONS
and
ESTIMATES
PREPARED.

MINING PLANTS
ERECTED ON
ANY PART
OF THE
FIELDS.

68 Boulder Road,
Kalgoorlie,

In October 1920, W.G.M. borrowed £50 by mortgaging the office premises at 68 Boulder Road, Kalgoorlie.

William Edwards, Auctioneer & Estate Agent arranged the mortgage at 15% interest.

Interest was paid regularly until W.G.M. became ill in 1924.

In May 1924 Edwards suggested a settlement of £73-9-4. This was paid on August 26 by C.B.M.

Margaret Manners at 7 Croesus Street, Kalgoorlie - 1925.

The woman and I got into conversation and, of course, could not long avoid the subject of politics. I mentioned that someone who passed at the time was a good fellow, but a rotten Sinn Feiner. This raised her ire and she, metaphorically speaking, flew at me. Then the boy lent his assistance to the argument. His remarks were.

"I hate England. I hate everything English. I'll fight England. I'll fight against England every chance I get."

Here was a child who probably did not know where England was. I am sure he did not know its history or traditions. In fact he knew nothing about it at all except the one great fundamental idea that had been fed to him with his mother's milk and had been the foundation of his education ever since. It had been in his prayers. The reason for his hate was beside the question. He could not have given one single reason for his hate; all he knew, all he had been taught, was "I hate England."

Note: By C.B.M. (Approximately 1960) and updated by R.B.M. (1991)

This chapter was never completed as the Author passed away after suffering patiently with cancer. He was a man of sterling qualities whose life enriched the lives of all who knew him. Having shown us how to live he surely showed us how to die.

So much of his life story goes unrecorded for in the words of Burns, whom he so fondly quoted *"the best laid plans of Mice and Men..."*

It seems that some fragments of his great talent and experience may be drawn together in conclusion of his own writings and it is with humble apology that I will make some endeavour to accomplish this.

My father's gift of writing was such that the whole of the preceeding twenty nine chapters were in the original script and sequence, having been composed in the last few months of his life as they are finished here. There is one exception and that is the inclusion of his poems and letters. (Gathered together in Appendices I & II).

In 1935 correspondence was received from the ninety four year old James Lamb of Adelaide. It has such a bearing on early Broken Hill and MacDonnell hopes, that it has been included in Appendix III.

CONCLUSION & ACKNOWLEDGEMENTS

Following the initial writing of this material by my grandfather (W.G.M.), and the later assembly by my father (C.B.M.), it was left for me to complete this project and to track down photographs, maps and illustrations, as W.G.M. had an office fire just after World War I and lost all his records. My father died in 1966 before completing this project which he regarded as a final tribute to his own father, whom he certainly admired greatly.

To all those people in Ballarat, Broken Hill, Townsville, Melbourne and Kalgoorlie who have responded in detail and with patience to my many queries, go my sincere appreciation.

My thanks also go to my sister Frances, my wife Jenny, secretary Fiona, Peter Bridge, Hal Colebatch and Audrey Webb for their proof-reading, suggestions and encouragement.

Local historians Keith Quartermaine, Norma King and Ross Louthean deserve special thanks for their "nudging" throughout the project.

The book as it now stands, forms some early pictures of the beginning of Australia's mining industry and the challenge met by those early battlers.

I feel from the experience of preparing this material, that I have come to know my grandfather and I'm sure that I would have enjoyed meeting him personally and compiling his observations of our mining industry and our way of life in the 1980's.

In several sections of this book he high-lighted classic examples of the growing resentment against achievement. He observed this resentment from Government, from Labor Unions and from individuals.

Both resentment and achievement have co-existed throughout history, in all societies, although in widely varying degrees. When the morality of achievement predominates, civilisations flourish in commerce, in the arts, in sciences; and are remembered by future generations.

When the (im)morality of resentment gains the upper hand, civilisations decline and eventually perish. A civilisation is the sum total of all the achievements of its people, and as achievement becomes increasingly discouraged, scorned and even persecuted, the forward momentum of a society is quickly halted, and then ultimately reversed.

This (im)morality of resentment, or envy, is known in Australia as our "tall poppy syndrome". ie "Let's chop them all down to our size"

A growing number of Australians seem to be sharing my current concern, as to the direction our country is taking, and I am optimistic enough to think that there is a "turning of the tide" and that socialists after winning for the past seventy years, are seeing their vision fade. A recent poll of student attitudes in Australia showed that 74% support ideas of Libertarian free enterprise and 65% of those young people interviewed blamed governments or unions for our economic problems.

This is a reflection of what I experienced personally last year in the Soviet Union, where although the economic and political bankruptcy of their socialist-collectivist doctrine was obvious, it has surprised so many observers that the new "freedom uprising" is so effective and so widespread.

This phenomenon was possibly best described by President Vaclav Havel of Czechoslovakia, on New Year's Day, 1990 when he said:

"Throughout the world, people are surprised that the acquiescent, humiliated, sceptical Czechoslovak people, who apparently no longer believed in anything, suddenly managed to find the enormous strength in the space of a few weeks to shake off the totalitarian system in a completely decent and peaceful way. We ourselves are also surprised at this, and we ask where the young people, in particular, who have never known any other system, find the source of their aspirations for truth, freedom of thought, political imagination, civic courage and civic foresight. How is it that their parents, the generation which was considered lost, also joined in with them? How is it even possible that so many people immediately grasped what had to be done, without needing anyone's advice or instruction?

Man is never merely a product of the world around him, he is capable of striving for something higher, no matter how systematically this ability is ground down...the humanistic and democratic traditions — which are often spoken about in such a hollow way — nonetheless lay dormant somewhere in the subconscious of our nations ...and were quietly passed on from one generation to the next in order for each of us to discover them within us when the time was right..."

From a similar new generation in Australia can be recruited the "Black Belt Libertarians" as I call them, people skilled in understanding and explaining how well the world has worked during the brief periods of free enterprise, and how well it can work again if we only remove the socialist shackles of both our major political parties.

I think my dear old grandfather would have enjoyed seeing these signs and he probably would have agreed with economist, Henry Hazlitt, who commented in his book *Economics In One Lesson*:

WORSE THAN THIEVES

"When your money is taken by a thief, you get nothing in return. When your money is taken through taxes to support needless bureaucrats, precisely the same situation exists. We are lucky indeed, if the needless bureaucrats are mere easy-going loafers. They are more likely today to be energetic reformers busily discouraging and disrupting production".

RON MANNERS (December 10, 1991)

FOOTNOTE: #1

The W.G. Manners & Co balance sheet reproduced on the following pages dated December 10 1924, (on the death of W.G.M.) shows trading stock of £115, sales of £279 and nett loss of £26, which certainly falls within the definition of "small business".

FOOTNOTE: #2

A start has already been made on documenting the next "generation" and it will cover my parents' experiences through to 1966. It is titled *Kalgoorlie - The Golden Years to 1966* By Charles & Nancy Manners.

OBITUARY

THE LATE MR. W. G. MANNERS

The mortal remains of the late Mr. William George Manners were laid to rest in the Presbyterian portion of the Kalgoorlie general cemetery on Friday afternoon in the presence of a large and representative gathering of citizens. The opening part of the burial service was said at the family residence, Croesus-street, by Rev. Mr. Saunders, of St. Andrew's Church, who also conducted the service at the graveside. Before reciting the closing prayer and pronouncing the Benediction Mr. Saunders took occasion to deliver an impressive address relative to the late Mr. Manners' sterling good qualities both as a man and a Christian. He had set a splendid example in home life and exerted a great influence upon all with whom he had been in contact, whilst in his public life he had always sought the betterment of the people with whom he lived. Though a sufferer of excruciating pain the late Mr. Manners had borne up against it with Christian fortitude, and was content to be at rest with the Eternal Father. The funeral procession was led by about 40 Boy Scouts of the St. Andrew's troop, an organisation in which he had displayed much interest. The principal mourners were:—Messrs. William M., Chas. B., Stanley and George S. Manners (sons); David H. Don (cousin), Mark A. Saunders (brother-in-law) and Beverley Saunders (nephew). The Breadalbane Lodge, No. 80, of the West Australian Constitution of Freemasons was officially represented by Messrs. David Sheed and Duncan Macgregor and other members of the craft. The Mayor (Mr. F. W. Allsop), Cr. H. L. Barnard and R. G. Shaw (town treasurer) attended on behalf of the Kalgoorlie Municipal Council, of which the deceased gentleman had been at one time a member.

Mr. John MacIver and Mr. F. C. Simpson were present for the Kalgoorlie Roads Board and Messrs. T. Butement, D. McDougall, and T. Knox and others for St. Andrew's congregation, and Mr. Harold Kingsbury (vice-president) attended to represent the Kalgoorlie sub-branch of the R.S.L. of which one of the deceased's sons was a member. The gathering also included Mr. Ernest Williams (Great Boulder Perseverance mine) and Ven. Archdeacon Brewis an old Murchison friend of the deceased gentleman. The pall-bearers were: Messrs. F. W. Allsop, D. MacMillan, D. McDougall, D. McDiarmid, D. Sheed, and D. Macgregor. Numerous floral tributes and messages of sympathy were received by the bereaved family. The wreaths included those from the Kalgoorlie Municipal Council, the Kalgoorlie Bowling Club, members and adherents of St. Andrew's Church, St. Andrew's Ladies' Guild, and St. Andrew's Girls' Club. The mortuary arrangement were carried out by Messrs. A. and J. Kyle.

The deceased gentleman was born in Ballarat, Victoria, 60 years ago, being the third son of the late Mr. William Manners, who was then a mine manager. After leaving school he became an apprentice to the Phoenix Foundry in his native city, and forthwith was enrolled as a student of mechanical engineerig at the Ballarat School of Mines. He was one of the first, if not the first, students to get the diploma of engineering at that institution. When he was 26 years old he married Margaret, fourth daughter of the late Mr. Charles Brown, of the Grenville Foundry, Ballarat. Shortly afterwards he was appointed to one of the Broken Hill mines, where he stayed seven years. When the news of the discoverv of Coolgardie and other goldfields was announced he came West and got the appointment of engineer for the Golden Crown mine at Kanowna. Staying there for about 12 months, he resigned in order to take up business for himself as a mining engineer and patents attorney.

He carried on those avocations up till 1911 and during that period he designed some of the plants on the Golden Mile, as well as in other parts of West Australia, being employed in their erection as well. In 1911 he was engaged by a London firm of mining engineers to proceed to Rhodesia and thence to London to design and erect a large plant for the Cam and Motor Mine, in Rhodesia. On its completion he returned to West Australia and continued to practice his profession up to about two years ago, when his health failed. He was compelled to relinquish any active part in the business, when he handed over the management to his son, Charles, under his supervision. The deceased was a man of many accomplishments, and took keen interest in politics and was well read in political economy. He was a member of the Kalgoorlie Municipal Council, and resigned on account of ill-health. He was a good public speaker and debater and took great interest in young people, and was a great lover of clean sport.

The deceased leaves a widow, five sons and one daughter to mourn their loss. They are all residents of West Australia.

Kalgoorlie Miner
December 15 1924

Margaret with W.G.M.
August 1924

The death of ex-Councillor W. G. Manners was referred to in a sympathetic manner by the Mayor (Mr. F. W. Allsop)) at the Kalgoorlie Municipal Council meeting last evening. The late Mr. Manners, who severed his connection with the council two years ago, had since passed away after a long illness. He was loved and respected by all who knew him, and he possessed one of the best and kindest dispositions. It could truly be said he was one of nature's gentlemen. The Mayor concluded by moving that the council convey its sincerest sympathy to Mrs. Manners and family in their bereavement, and that a minute be placed on the records appreciative of the services ex-Cr. Manners had rendered to the municipality. Cr. A. C. Wall, who sconded the motion, testified to the regret he felt at the loss of a worthy citizen. Cr. H. L. Barnard supported the motion. He had known the late ex-councillor for 35 or 40 years. They first became acquainted when the late Mr. Manners was a student of the Ballarat School of Mines, when he (Cr. Barnard) was assistant registrar of that institution. The resolution was carried, councillors standing in respectful silence.

Dr. Office Copy

W. G. Manners

Profit & Loss Account — 1st July

	£	s	d	£	s	d
To. Stocks on Hand, 1st July	100	·	·			
Add. Purchases	114	7	10			
	214	7	10			
Less. Stocks on Hand 10th December	115	17	·	98	10	10
" Balance carried down				180	13	4
				279	4	2
" Wages.						
W. G. Manners	114	2	9			
C B Manners	130	·	·			
Sundry	46	12	6	290	15	3
" Freights - Cartages				11	3	5
" Travelling Expenses				31	16	9
" Rates, Municipal, six months	3	13	6			
Water	1	6	3	4	19	9
" Electric Lighting				1	1	
" Insurance, E.S. adjustment					4	4
" Stationery & Printing				4	7	5
" General Expenses				15	·	·
" Interest Paid, Est Stearne				2	7	6
				£361	16	1
				Supplementary		
To Balance bt down				30	6	5
" Bad Debts written Off				27	15	4
				£58	1	9

& Co.

to 10th December 1924 — Cr

	By Sales.	279	4	2
		279	4	2
	By Balance b/d down	180	13	4
	" Commissions Received	147	10	4
	" Patent Fees. Net.	3	6	.
	" Net Loss	30	6	5
		£361	16	1
1	By Commission Recd, Motor Car	31	12	9
	" Net Loss, amended	26	9	
		£58	1	9

W. G. Manners & Co.

Balance Sheet — as at 10th Dec 1924.

~~Sundry Creditors~~	825	8	1	~~825~~	~~8~~	~~1~~
Less. Balance owing on Motor Car included therein see contra	354	7	3	471	·	10
Capital						
W G Manners	97 ~~60~~	10 ~~5~~	~~10~~			
C B Manners	97 ~~60~~	9 ~~5~~	11 ~~10~~	194 ~~120~~	19 ~~11~~	11 ~~8~~
				£666 ~~£591~~	· ~~12~~	9 ~~6~~

Commonwealth Bank				73	4	8
Sundry Debtors				156	18	1
Stocks, trading				115	17	
Motor Car	474	8	3	~~400~~		
Less Balce owing, as per Contra	354	7	3	120	1	
Land & Building				200		
				£666	·	9

RECOMMENDED READING

Those Were The Days	-	Arthur Reid 1932
Glint of Gold	-	Malcolm Uren 1948
View from Kalgoorlie	-	Ted Mayman 1980
White Feather (The Story of Kanowna)	-	Margaret Bull 1987
Mines, Miners & Metallurgists in Western Australia	-	D.A. Cumming 1991
Voyage Of No Return (Manners and Relations)	-	Gladys E. Manners 1981
Kalgoorlie - The Golden Years to 1966	-	Charles & Nancy Manners (In Prep.)
Kanowna's Barrowman - James Balzano	-	George Compton & Ron Manners (In Prep.)
My Elusive Career	-	Ron Manners (In Prep.)
Mannerisms	-	Ron Manners 1983
Hoover's Millions And How He Made Them	-	James J. O'Brien 1932
Technological Survey Of The Golden Mile (2 Vols.)	-	M.K. Quartermaine 1981
Daughters Of Midas	-	Norma King 1988
The Mile That Midas Touched	-	Gavin Casey & Ted Mayman 1964
Broken Hill to Mount Isa	-	W.H. Corbould (Ed. Ian Hore-Lacy) 1981
Lucky City - The First Generation At Ballarat 1851-1901	-	Weston Bate 1978
Ballarat - Golden City	-	John Reid & John Chisholm 1991
Goldfields And Mineral Districts Of Victoria	-	R. Brough Smyth 1869
Broken Hill - A Pictorial History	-	R.H.B. Kearns

APPENDIX I

Early poems and notes by W.G.M. in Victoria

I also had a penchant for poetry in those days and was very proud of winning first prize for my poem in the year 1885. It was a most unpoetical flight of fancy, comparing our last competition to a stairway, something like Jacob's Ladder, leading to realms unknown above the stair shone 'Sols' illumined rays of knowledge. On the topmost step stood Avon's Bard with the lesser lights in relative locations lower down. Our competition aspirants were making attempts to climb these stairs.

"OUR LAST COMPETITION"

Pause for a moment where the mind aspires
And views its valiant flight on Fancy's stair
The world a platform where a surging sea
Of souls untutored swaying to and fro
In aimless pathways seek immortal name
Out from the midst of this meandering mass
A flight of stairs arise, where mortals oft
Aspiring, mount to realms of laud and fame.

Upon the summit gleams a golden orb
From which fair affluence in a genial shower
Pours down its radiant rays upon those souls
who strive untired, approach towards its glow
And for their labors claim a rich reward
Of guilded honour. As does fair Lima
From the rays of Sol reflect a light both
Luminous and lustrous for the use of human kind.

So doth the aspirant, receiving beams of glory
from the Orb, recast them to the living world beneath
Which give a lasting boon to all creation.
In such a flight as this our contests guide
The great ambitions of too frail mankind
And sway with satisfaction to the throne
Of fairest fame the meliorating mind.

The fallen curtain of departed days
Has veiled from view Lore's last contesting scene
But they alone have vanished from the view
For still the influence of the famous fray

Is budding and in bloom, as on the flowers
Delay the favors of the setting sun
Till morn awakes refurnishing his charge
But raise the curtain and disclose these days
Imprint them on the memory of the world
As days of wondrous virtue and divine
Then see the flight of stairs the shrine of all
Firm from the floor arise beginning youth
Allured to fields of fame by fellowmen
Who forward toil. One step possessed makes strong
The true ambition for a higher claim,
And each attainment makes foundations firm
For future aims extending to the skies,
As emulative minds will ever cling
With unremitting zeal to such a state.

How sweet the pleasure that success supplies.
How grand and strong that power magnetic is
That once its influence felt the mind must fly
To perfect state, for with each ruddy glow
Desires shall dawn facilitating fame.

There on the topmost stair stands Avon's Bard.
His colleagues cluster round. Tis they that claim
The foremost glow of Glory's Orb and shed
In all the eyes of all the world its fair
Reflection. While beneath in right rotation
Or deserving deeds, Creation's sages stand
Receiving and reflecting as their lore
Allows, the gleams of fulgent light
Which guild them that they brilliantly appear
Creation's gems. On such a stair as this
Imagination views a heaven of stars
Beaming with beauty, brilliancy and light,
The Admiration of the universe.

All by the same great orb of fame illumined.
This Orb is ours. The last display of lore
Extended wide the great essential boons
Of swelling fame revealing latent power.
Alluring to its light full many a mind
Imparting knowledge round which will betimes
Gleam out with peace and power on nature's plan
And manifest the best of better Man.

In the same year I wrote the following lines about our own land.

"AUSTRALIA"

Fair Austral! Monarch of the southern wave,
Whose rock-bound coast the surging oceans lave.
And sheltered inlets, formed to nature's plan,
Display a shore invitingly to man;
Thence wondering navigators strained their eyes,
But every gaze created more surprise;
So vast, so beautiful, so rich a shore,
Inspiring zeal its inland to explore,
Through distant plains and thickly wooded lands,
Uncultured by the art of human hands;
Where savage tribes with glad but hideous yell
Proclaimed that freedom which they loved so well.
With spear and club they hunt the kangaroo,
And chase the dingo, or the swift emu
O'er the landscapes wild, yet decked in beauteous grace,
For e'en their wildness ornaments the place.
And distant hills, though but to vision known,
Seemed all the world exclusively their own.
Oh! Praise them then who first all fear dispelled
And entered this dark land by hope impelled.
Who first discovered and in Britain's Isle
Proclaimed its golden charms and for the while
Inciting British courage to aspire
To reach that land and happiness acquire.
Strong ships were laden with a human freight
Responsive to the call of "Emigrate".
For Fancy's visions of the future gleam
Before their eyes and form a constant dream;
Until by gold's sweet fascination led,
They, far across the rolling billows sped.
E'en as our hills when mantled o'er with snow,
With sparkling grace their pure adornments glow,
Till envious Sol his ruling powers regain
And rob the joys they would at first retain;
Then quickly parting with their white array
Accept his golden rays more bright than they,
So 'tis with man though much attached to home,
Allured by brighter hopes he longs to roam;
They, leaving comforts all; on wealth intent
Adopt the humbler fare a canvas tent:
So in they pour, till every sylvan creek
With noisy tumult, did its story speak;
The cradle's roar, the rattling dish's song
The shouts and clamour of the busy throng:

And stalwart miners, penetrating ground,
In searching deep the soil where gold is found,
Hail golden visions, with glad acclaim,
Of riches, splendour, happiness and fame:
And yet how real; these tents have been effaced.
On their foundations stately homes are placed
Then towns and cities through the land arise,
Embracing all that tends to civilise
And gem her plains, as flowers the fields improve,
Or stars bedeck their azure bed above.
So art and nature joining hand in hand
With mutual purpose beautify the land;
Till now the sun resplendently patrols
And spreads his radiance on three million souls,
Who onward strive by industry and toil
T'unearth her riches, cultivate her soil,
Open her lands where nature's secret stores
Disclose a field of flora and of ores.
There ancient men whose lingering lives delay,
Stand proudly in the dusk of parting day,
And point with joy to yonder lofty spire
Of holy church, which points to object higher;
Or yon asylum in the suburb seen
In sheltered nook, embowered in evergreen,
Where ageds and indigents, in comfort live,
To swell the happiness of those who give;
Or gardens rare by nature doubly blest,
Find zealous man productive of the rest;
E'en as a prince who nobly born and bred,
In youth, has honours lavished on his head
Although proclaimed his own, yet truly, he
Received them, from his noble ancestry;
But when he, to maturer years has grown,
Must earn a reputation of his own.
So thee, Australia, though thou'rt now proclaimed
A glorious land for wealth and beauty famed,
Yet time will add by self achievement, quite,
Arts, science, letters, virtue, courage, might;
These will pervade thy vast and verdant plains,
Thy tow'ring hills where virgin woodland reigns.
And oft a deep impenetrable grove,
Or rippling streams which through the valleys rove,
Or lovely lakes in silv'ry charms arrayed
Where fragrant verdures form a margin shade
And in their mirror beds reflected, lies
The sapphire tinted concave of the skies,
Then here and there a hamlet intervenes
To mark Aurifera's contested scenes;
And all abroad such charms as these abide

In artificial and primeval pride.
Again it seems thy great creators will
To render these adornments brighter still,
By making thy inheritance a clime,
In all thy countries pleasantly sublime:
Far in the north where, scarce obliquely, fall
Sol's gleaming rays on palms erect and tall
Which far and wide in rich redundance grow,
And mankind's best necessities bestow;
Their fan formed leaves the sheltering thatch afford
Their fibres clothing, fruits a sumptuous board.
In New South Wales awhile the vision roves
Among the vineyards and the orange groves,
Which gild the orchards, charm the artists' eye
And every homely luxury supply.
Then south again to where Victoria's soil
Remunerates the hardy sons of toil
Where Art and Science all predominate
To cheer the thrifty swains who cultivate.
In short, Australia is from shore to shore,
Proclaimed by experts in Botanic lore,
To be, regardless of her recent birth,
The great, unrivalled garden of the earth.
Her flowers, her fruits, her crops, are unsurpassed,
Besides her mines so varied and so vast.
While great successes with the "golden fleece"
Have been achieved and promise to increase,
We laud Macarthur's far, perceiving eye,
Who first discerned and did that want supply,
While gold for others held magnetic charms,
He brought the flocks which decorate her farms.
Her merchants too with unremitting zeal
Through all the world in her productions deal
That with her wools and wines comperes defy,
In quality and plenteous supply;
And each domain with emulative aim
Contests their power, their products and their game,
While each ambitious to excel the rest,
Of good makes better, and of better best,
Our thoughts now turn with sympathetic sway,
Where famished Bourke and Wills in honour lay.
Oh! noble band, true patriots of the State!
Can man look down upon thy fearful fate,
Without a tear, a sigh, without a moan?
No, not when thy sad history is known.
Thy friends today have planted many a pile
That future ages on thy name may smile,
And that posterity may sing thy praise
In songs of pride and sweet pathetic lays.

We see again on recreations field
The progress made by Austral's sons revealed:
Where Murdoch's bat and dauntless Beach's oar
Repeat the claim so often claimed before.
Advance Australia! May thou never cease.
Assume the eminence of ancient Greece,
And recreate your own Demosthenes,
Your Alexander and your Socrates.
Create your patriots true, and dauntlessly
Maintain your honour and your liberty.
That liberty which, thirty years ago,
Our noble fathers earned with desp'rate blow;
When British soldiers drew the fatal blade
'Gainst Britons at Eureka's fell stockade,
Where miners justly held rebellious hand
Against an unjust, tyrannous command.
Oh! fearful sight! the angry miners wait
All armed and ready to stand, to meet their fate.
'Twas Sunday morning, cloudless was the sky,
And glowing nature watched with pitying eye.
December's zephyrs whispering through the trees
Try soothingly, their anger to appease.
But all without avail, for now in view,
Advance the daring, though reluctant few.
"Advance." they cry. "Charge soldiers, fight, defend!"
Then rattling shots and human cries ascend
To heaven's high arch. Then back the echoes dart
And pierce to core, each sympathising heart.
The miners shout. "Destroy your servile ties.
Demand that liberty the law denies".
They fought: were conquered: but their object gained,
Their first emancipation was obtained;
Which opened quickly longed for freedom's gate,
And gave self-legislation to the State,
With universal suffrage for our lot,
To rich or poor in castle or in cot,
And vote by ballot, great and sacred right,
That gives the weak the strength to cope with Might.
Thus they resisted for a rightful claim,
And who dares say "Disloyalty their aim"?
What realm in all the world has ever known
More loyal subjects than sweet Austral's own?
When Federation triumphs in the land,
Then will her sons and daughters, hand in hand,
With heart and voice, o'er hills and valleys green
Resound the loyal song "God save the Queen".

In 1886 ***Australian Federation*** was the title of another writing:-

Dame Austral sat beneath her hall of state
The Canopy whose gems resplendant gleam
In Sapphire beauty, while her daughters wait
To hear her words of lore, which could redeem
Those dying favors, that a feudal dream
Have vanished for a dawning doleful wrong
Of emulative strife, which now would seem
So dire, immovable, alas, too strong
To learn to wisdom now, that should to peace belong.

Ye Sisters of the Southern Hemisphere,
My daughters fair, give audience awhile,
That peace elate which now thy homes endear,
E'er summer suns have shone, may cease to smile.
Then sweet accord abounding in our isle
Would swell our power, while selfishness will shine
The false mirage that would our hopes beguile.
As sylvan streams to one their courses twine
Create a mighty river bounding to the brine.

*List! Fair Colonia**, *thou rebellious Maid,*
Why, still, thy sisters grand presentment waive?
Must it be so if harmony pervade
This golden land? thou failst alone to save
This grand Ideal from a lonely grave.
Have emulations, hopes, thy love erased
From home and peace? Ye seas such barriers lave
That siren songster is for ruin praised.
Join hands, join hands, ye realms, and leave the world amazed.

See Heavenly concord round the ruling Sol
Ten thousand golden worlds subservient roll.
Each aid are aided till harmonious all
The great divine supremacy extol
And manifest perfection of control.
Oh! thou frail man these wonders overcome
Reveal the frailty of the human soul
Thy ruling powers inanimate, aye, dumb,
And fades to future view our grand millennium.

It ne'er can be where man imperfect reigns.
This rival separation must pervade.
For foreign powers, ambitious of domain,
Meanwhile molest, while all the world dismayed,
But thou fair daughters, why shouldst thou upbraid
When peace abundant is thy living dower.
Avoid this havoc man to man has made,

Let unity of aim thy land embower,
View heaven's domain complete and
Emulate her power.

Begin Fair Maids, give separation wings;
Were concord thine 'twere thine for evermore.
Could earth retain a thousand rival Kings,
'Twere but to make a thousand Kings deplore.
Speed on, good Pegasus, Creation o'er
Thy potent song proclaim, Link man to man
Minerva's voice I hear "Peace, peace restore,
We cannot end the World as it began,
It shall be pure betimes and heaven its purer plan."

* *N.S.W.*

I was not ambitious to be a poet, although rhyming was quite easy to me. I had figured that all poets were poor and unappreciated until they were dead, and that was no good to me. So I contented myself by writing satirical skits on current events such as *The Parson's Man*, or as it was later titled ***Harry's Fate***. This was founded on fact.

H. Little was a Parson's Man
Who early learned to prize
The path directed by his nose
Which pointed to the skies.

But Harry, spite of christian grace,
Was prone to human bliss,
And deeming he could not mistake
Resolved to take a Miss.

He straightway courted Susie Lee
And though he was no flirt,
He also courted rather long
And Susie, she felt hurt.

To have the trouble overcome
She very quickly planned
That as they stood upon the lawn,
She'd make him understand.

So when at evening they communed
She vowed 'George Love' had said
'She was a perfect flower', and wished
She'd share his daily bread.

But he, perceiving not the joke,
Ne'er dreamed of honeymoon,
He only cursed his luckless stars
And then thought fit to swoon.

Then straight into a chemist's shop
For poison Harry ran
And vowed 'ere close of day to close
The days of parson's man.

When nose and woes alike turn up
To lips the goblet goes
And first he turned the goblet up
And then turned up his toes.

When first the parson heard the tale,
A sorry man was he,
He had not only lost his man
He lost a marriage fee.

But Suie, when she heard his fate
"Oh! Wan am I!" she cried,
"Since we two never can be one
I will die too", she died.

Now, as the cause that killed them both,
Was to one issue traced,
And death their only joy could earn
They in one urn were placed.

They wrote upon the marble tomb
"Here lies a Suicide.
He dies for Sue and that is why
He lies by Susie's side".

Among other pieces written those days were the following

"THE MAGIC OF A KISS"

A mind untaught in muse's lore,
That Love imparting bliss,
Can never fully comprehend
The magic of a kiss.
To them it has no form but sound,
No tie save friendship's kind,
To them it plays no other part
Than please a friendly mind.

But when in Love's deluded ear
Its fondest music rings,
All earthly joys are vacant while
Its lingering echo sings.
The hands might join in friendly clasp
Or noses rub again,
A smart young man may raise his hat,
Or wave his walking cane.

A larrikin may wink his eye,
A gent may bow in pride,
A Lady's maid may charm us with
A curtsey low and wide.
But what are these but trifling things
Devoid of even bliss,
Great Kino of Salutations is
That loving little kiss.

They stand beside the gate and talk
Of love, and loves reward,
Their eyes for ever meet and hearts
Beat praise with one accord.
Their hands are clasped and by the power
Of Electricity,
From heart to heart each beat conveys
Its praises earnestly.

The time has come to bid farewell,
That word so hard to say,
His eyes request while hers invite
Then shyly turn away:
He clasps her neck, his eyes grow dim,
Mind is a moment lack
Their lips are pressed a moment, then
Oh! heavenly pleasure "Smack!"

'Tis done 'Tis o'er, No, 'tis not o'er
Its music has dispersed,
Though often in the memory is
The sweetest sound rehearsed,
It speaks what tongues desire to say,
But oft attempt in vain,
It binds fond hearts in sympathy,
And tells of parting pain.

THE BANKS 0. YARROWIE

When waning Luna's oval orb
Had wandered in the west,
And evenings placid music lulled
The weary swains to rest.
The quickening hours flew on alive
With unremitting glee
And mirthful moments merry made
The Banks 0' Yarrowie.

The timepiece told the fleeting hour
And bade us all away,
The laugh bespoke the happy hearts
That lengthened our delay.
The sweet magnetic chains of love
Some born and some to be,
Enhanced the smiles that e'er adorned
The Banks 0' Yarrowie.

FRIENDSHIP

14.1.1885 To Miss Maggie Brown [Later to be Mrs. Manners]

How sweet the sensations that friendship imparts,
No sweeter our Guide has designed,
They e'er have a permanent place in the heart,
And a prominent place in the mind.

And when like the beauties of spring they appear,
And bestow all the happiness known,
They enliven the fast fleeting moments with cheer,
With the thoughts of the days that have flown.

E'en thy presence or vision my musing inspires,
And allures sweet impulses to view.
Then I deem that a heart little other requires,
Than to cherish a friend who is true.

'Tis my wish that thy future may ever remain,
Wheresoever thy lot may be cast,
As peaceful, as happy, as void of all pain,
As the jovial days that are past.

When your life's passing days to the gloaming has drawn,
May its darkness undoubtedly prove,
But the shadow of evening preceding the dawn,
Of the shadowless morning above.

OH! WHY SHOULD THE HEART OF A MORTAL BE SAD?

Oh! Why should the heart of a mortal be sad
With love for its only desire?
Or can true love reign in a heart that has had
A tendency ever to tire?

Oh! Love thou art strange, Thou art strong as a chain,
And as tender as tendrils of vine;
Thou wilt bear the affliction of tempests, while rain,
Gentle rain, will cause thee to repine.

As a pebble that lies on the bed of a bourne
Will cause the smooth surface to surge;
Yet not a great mount could those waters return
To the fount whence at first they emerge.

So Love, but the lisp of a lover may send
All his hopes to a fickle dismay,
Yet no power ever known can ever pretend
To decoy that affection away.

Had I known that to lavish my love could offend,
To admire thee alone I'd relent,
With so light a beginning, so heavy an end,
E'en the shadow of love should prevent.

Oh! why should the heart of a mortal be sad,
To Love's own devices entrusted?
It is when those ties are all severed and may
Be never so firmly adjusted.

When Love the Heart's daughter is lain with the dead,
And the Heart all in mourning is clad,
And all that it beats for, for ever has fled,
Then license a heart to be sad.

"ODE TO LIFE"

As Life's gay stream goes rolling on,
It meets with many changes,
It glides through valleys, plains and falls
And rough and rugged ranges.

Though pearls and rubies deck its bed
And flowers grow round redundant,
Yet thorns and reptiles bane and dread
And rocks are more abundant.

KINDNESS

When one with genial nature blest,
Bestows a kindly action,
Its own resulting happiness
Is ample satisfaction.

But if a gift reward be made
Its joy away is driven,
It robs it of its heartiness
And makes it bought, not given.

TO A - B - ON HER BIRTHDAY

Though artful muses may compose
In rural rhyme or princely prose
The fervent wish that friendship shows
And sing it well
They're not the sentiments of those
That in us dwell.

This card doth wish the happiness
Thy natal day this day to bless,
Now tell me could I wish thee less
Or be more tame?
No friendships prompted earnestness
has fanned the flame.

My wishes being more sincere
Would give thee now unbounded cheer
Which would be thine from year to year
By woe unswayed,
That every June becomes more dear,
More richly laid.

And as the blossom on the rose
By each new spring its beauty grows
And beauteous bloom abundant blows
In perfect leaf
The while its aged drooping shows
no sign of grief.

May thou each natal day assume
As years roll on a richer bloom
And every pleasure to thee come
In plenteous showers,
That e'en thy pathway to the tomb
Be strewn with flowers.

THE FLOWERS

Only a few fair flowerets
In modest beauty planned,
Yet decked in sweet arrangement
As if by Flora's hand.

I love them for their beauty
Their fragrances that flow
Yet how they speak of other realms
Where their companions grow.

They guide imagination's eye
Far in the forest green
And fade, it seems, the hills that form
A barrier between.

They talk of lands where wild heath fumes
And morning breezes blend,
Where many happy memories dwell
And shadow thoughts attend.

There was a "Poets' Corner" published every Saturday in a supplement to the *Ballarat Courier*, and many of my friends tried to persuade me to publish something. But it was generally such awful tripe that appeared, that I did not feel like joining the Poetasters. There was a man called Brown, who was a regular contributor. He signed himself "D.A.B." I have none of his alleged poems, but this was about his style -

Man is a creature of the Universe,
But what a small thing is he,
He lives a little while alas!
And then is gone, you see.

Then there was Mona Marie, who wrote Epitaphs. Anyone who 'passed out' during the week was sure of her attention. No matter if he was a veritable Jack the Ripper, he was always sent to heaven.

Alas! John Smith we grieve for thee,
for Heavenward thou hast flown,
Leaving a wife and children three
To mourn their loved, their own.

I think she should have spelled her name "Moana", as all her poems were dirge. These were among the best of the Poets' Corner contributors.

However, a little incident forced me into this Corner one day.

D.A.B. had two sisters, Alice and Annie. Annie was very plain, but a fair actress who made a good Mrs Malaprop in our scenes from *The Rivals*. But Alice was different. She was romantic, one of those sweet things who love to be mauled and kissed. I had taken her home one night. She led me in to the back gate, down a dark lane, and we stayed sometime at the gate saying goodbye. This type was so distasteful to me that I satirised the episode in a rhyme. Any sensible girl would have seen the satire and have either torn it up, or hid it away in her private box. Not so Alice, she prized it, and took it as the real expression of my feelings. It was titled :-

WOOING IN THE LANE

They talk of lovely Wendouree,
With beautiful parades,
Its rural verdure spread around
Affording peaceful shades.
Its quiet nooks, where lovers meet
Their mission to explain;
But none of these are half so sweet
As wooing in the lane.

When stars are twinkling in the sky,
Their bed of modest blue,
And nature gently fanning scarce
Disturbs the falling dew.
The lingering silver light, while yet
The moon is on the wane,
But adds perfection to the scene
While wooing in the lane.

I brought a rose and lovingly
It to her bosom pinned,
And stole a kiss, alas, no crime,
Who ne'er stole more ne'er sinned;
My eyes looked softly into hers,
And hers to mine again.
The happiest days in life were then,
While wooing in the lane.

Her waist I clasped with pressure light,
Her little hand in mine;
While pillowed on my shoulder did
Her head in love recline;
The words she uttered seemed to bear
A sweet poetic strain;
'Twas charming music in my ear,
While wooing in the lane.

How sweet to dream of bygone hours,
Alas, fled far away,
Except to memory, where they are
As fresh as yesterday.
Oh, for some power by which we may
Such happy hours retain
Such memories never should decay,
As wooing in the lane.

My mind engaged in worldly strife
For ever shall return
To those, the happiest days of life
When I to love did learn.
And though all thoughts may vanish, yet
Alone will there remain,
Those days I never can forget,
When wooing in the lane.

Shortly after this we had a picnic to the Gong Gong. I paid no one any particular attention, but enjoyed the fun with all. In the evening we had a social, and Alice and Annie, in order to captivate were present, decked out with a liberal supply of 'Pearl' cream.

This was particularly abhorrent to me, I resented it. But another reason was that I met another girl there, (who attracted me so much that she afterwards became my wife.) Consequently, Alice and Annie were neglected, and evidently felt hurt, so you can imagine my chagrin when I saw in the Poets' Corner on the following Saturday morning my rhyme Wooing in the Lane over the signature "W.G.M."

Oh! what a time I got in the shop that day from my shopmates.

"I say W.G., what were you doing in the lane last night? Who was the tart you had in the lane, old chap?" etc. etc. At home it was not much better, I was the butt of all. I knew it had been done for spite, so I sought revenge by writing and publishing the following, under the title :-

"THE DECEPTION"

Oh! waud some power the giftie gie us,
Tae see oursels as ithers see us."

Burns

On last Guid Friday night, I strayed
Doon tae a social; where each maid
Was in her brawest goon displayed,
An' sweetest face,
An' happiness, awhile delayed
Aroun' the place.

Ye'll ken, I lov a' maidens fair,
What will wi' Venus' grace compare,
An' hae o' modesty their share,
That grace to bless;
For that, I'd lov if naething mair,
Or naething less.

An' sae I gloured aroun' the room,
Tae fin' the lass that could assume
The sweetest smile; an' bear the bloom
O' nature's art.
An' soon I spied the ane tae whom
I'd gae my heart.

Her neck and face were lily white,
Her cheeks eclipsed the roses quite,
An' as the rainbow's colors bright
Sae sweetly blend,
The rose and lily did unite
Upon ma friend.

Sae af I gaed, tae where the dear
Did 'neath the gaslight rays appear,
The richest of the rich, in cheer,
And beauty's wealth,
An' booing low, did kindly spier
About her health.

But as ma e'e her haffets met,
The sight I never can forget;
For ne'er did I, before nor yet,
Become acquaint
Wi' any lass sae much in debt
Tae Barry's paint.

I thought I'd waukened frae a dream,
As nae deception did I deem;
For she did sic an angel seem
A while ago;
But, ah! 'twas a' that fraud pearl cream
Which made her so.

Oh! vain, deceitful, foolish chiel
That would tae art's false aid appeal,
Tae nature's frailties conceal,
Tho' frailties pure.
If painters on the rose would deal,
They'd spile the fleur.

Noo a' ye maids wha dae the same,
Abstain from sic a foolish game,
For lads will only love the dame
In nat'ral state.
Deceit turns modesty tae shame,
An' love tae hate.

This hit the mark for next week appeared a reply from "Daisy", in the following strain. It is evident that she had also noticed the lady in question

"IN REPLY TO W.G.M."

Sir, I was at that social, too,
And saw the maid described by you,
I saw her cheeks, with blushes red,
But not the vile compound you said.

'Twas Nature's paint her cheeks adorn'd,
And Nature's hand her being formed;
Then if thou must satiric be,
Have Nature for thy enemy.

And when thou dealest in satire,
Light it with original fire;
Do not steal from other's fuel
To make thy spleen more fierce and cruel.

And Oh! Be sure the truth to speak,
Else all thy satire comes to grief;
And Barry's paint and the pearl cream
Makes thee contemptibly mean.

And "Manners should indulgence claim"
Should be our rule in joy or pain,
Some use them - as their lives proclaim;
While others use them but in name.

Had you been taught your manners right,
Your private spleen that social night
Had never come to public light,
With all its falsehood shining bright.

This effort was evidently from the pen of "D.A.B." because no one else would rhyme formed with adorned, or proclaim with pain. Besides, the meter of the line "Light it with original fire," was true to type, so I replied to "Innocent" Daisy as hereunder :-

REPLY TO THE "INNOCENT" DAISY

Now "Daisy" my friend, I did not intend
To cause such uncalled for confusion;
Nor would I have known, unless you did own,
To you I was making allusion.

To fashion at large I levelled my charge,
To moralise, so I intended;
But this moral lies in your moral eyes,
As if more ill lies had offended.

Beware, Daisy fair, if you wear such ware
Do not to the public declare it;
You're plainly aware my victim you were,
So if the cap fits you can wear it.

To aid my satire, I stole from the fire
The fuel of Burns, so you said it,
But stealing 'twas not, for flames that I got
I gave him the light and the credit.

If ladies I blame would just do the same,
And own their adorners, - Nay, tarry,
It would not look well, if each rosy belle
Would write on her cheeks the word "Barry".

If I may aspire I'd like to enquire
(As I have a narrow conception),
In blank verse or rhyme did you write last time,
Or with both did it bear connection?

But one query more, but one I implore,
Did "Dave" write it all, or you with him;
He'd greatly improve if nonsense he'd more,
And study more closely his rhyme.

On the Saturday evening when this appeared, Alice and Annie accosted me with a desire to know whether I referred to their brother Dave in the paper.

"No", I replied. "It is Daisy's brother I refer to. Are you Daisy?"

After some talk and mutual recriminations we became friends again, but there was no more wooing in the lane.

These Associations did a lot of good in many ways and were the means of bringing many men into the light of day. We had a lot of fun out of them as well, and it was artists from these that raised the money for the Statues of Shakespeare, Burns, and Moore, that adorn the main street of Ballarat.

APPENDIX II

A selection of W.G.M.'s poems written on board ship or on early arrival in the West.

THE THOUGHT THAT STRIKES THE ROVER (1896)

There's a thought that strikes the rover
Who's been tramping through the bush,
When the rush for gold is over
And he mingles with the push.
Who have been around prospecting
And have sold their little show
And are blowing it expecting
They will have another go,
When the boozing and the ladies
Have sent all their cash to Hades.

And the thought that strikes the rover,
Is that pluck and toil and graft
Has its little day of clover,
When it drank and sang and laughed.
Then he takes a job as trucker,
In his wallet - not a cent.
So he has to earn his tucker
And assist to pay the rent
Of the Landlord and the ladies
Who send all his cash to Hades.

"WHY THEY'RE SILENT"

(Westralia 1896)

All our pals have been successful
In the glorious Golden West.
We should have made the venture,
And succeeded like the rest.
Look at Jones and Brown and Melville,
Who at Christmas time did shine,
When their money flowed like water
On the effervescing wine.
Ah! The way they raised our envy
With their grandeur and their gold.
Yes! We should have made the venture,
We should have been more bold.

This erroneous impression,
Generating wild dismay,
Was the source of all the trouble
With the swain who stayed away.
"Ah! They all have been successful."
Poor human nature's wail
Who remember fools who flourish
And forget the fools who fail.
For the lucky are so careful
To return with lavish noise
And proclaim it to the chagrin
Of the stay-with-mother boys.

Have they all been so successful
Who have braved the arid West?
The thousands who are silent
Tell the gruesome story best.
You remember, say a dozen.
What of those you have forgot
Ah! a hillock in the desert
Marks their claim and final lot.
There are thousands who have fallen,
There are more who never rose,
And are consequently silent
To the world about their woes.

It is only the successful
Who can well afford to boast.
That ever come to tell you
And invite you to a toast.
Oh! the silent ones, the number
As they toil for drink and food.
Who can count them, Know their troubles
Or assist them if he would?
For they wander in the desert,
Or they loiter in the town,
And a train of circumstances
Always seem to bear them down.

Do you wonder why they're silent
Or remain across the deep?
While your friends return to show you
All their grandeur while you weep.
Your friends? You have forgotten
Some you loved, the very best,
And they died for want of water
In the golden starving West.
Or they wander and are toiling
Or are out upon a spree.
But they're very, very silent
As they have good need to be.

DAY DREAMS

(1895)

(Written in S. Australia on my way to West Australia)

How sad are the themes of the lover when banished
By forces of fate from the fair, whom he loves.
When the vision is near but the bosom has vanished
And the dream of the past but the present reproves.
Still sweet is the sadness that swells the emotion
And fills to o'erflowing the gates of the soul,
As we dream of the tokens of parting devotion
And the smothering sobs that we failed to control.

Oh! Why should the heaven-bound soul set asunder,
When God has ordained we are living as one?
It is man fighting man in the struggle for plunder,
When the feeble must fail ere the battle's begun.
O'er the land wide extending we follow the vision
Of Fortune's Mirage, o'er the desert of hope,
Who flees our approach with alluring derision.
As, with love for implusion, with failure we cope.

Let the struggle be fierce, for ambition is fearless,
And the haven in view is a princely reward.
The hope that impell though the prospect be cheerless,
Gives power to the arm and an edge to the sword.
The cares of the past we no longer remember,
There is nothing remains but the beauties of old.
The alluring, inspiriting glow of the ember
That promises pleasures in days yet untold.

Sweet delightful delusions our present beguiling,
The fate of the future our dreams have forecast.
The flowers in our garden of life that were smiling
Have founded these fancies on joys of the past.
These flowers that have bloomed long ago and faded.
The flowers, everlasting, of love that remain
Have scattered the seeds and are budding as they did
And promising pleasures exceeding again.

And, while dreaming of days that are gone, how we ponder
On favours of fate that our visions foresee,
For the fond banished heart will grow fervently fonder
And flies in its fancy far over the sea,
Where the dear ones are waiting and loving and longing,
And praying for those who have gone to return.
While the fears are intruding, dark images thronging
And dreamings of doubt in the bosom will burn.

Though our hopes are but human, devine in the sequel.
And Oh! may our hopes and the sequel agree
They are, but that the days that will come, may be equal
With those of the past. They were heaven to me.
Oh! Sweet are the themes of the past and, still sweetly,
The dreams of the future no shadow forecast.
We live in the present on hopes that completely
Devine future joys by the joys of the past.

WINK THE OTHER EYE

(1895)
(Written on the S.S. *Barcoo*)

When the restless waters roll,
Which the shifting winds control.
With the cherub of my soul,
My Lizzie dear and I.
Each other, fondly fold,
We cannot feel the cold,
For love is warm, I'm told
As I wink the other eye.

She sails to meet her beau,
Fondest love to e'er bestow.
He will never see or know
How Lizzie dear and I
Spent the time in sweet embrace.
Nor will he ever trace
All the kisses on her face,
As we wink the other eye.

Eyes that see not, cannot grieve,
Hearts that love, will not believe.
So we spend the dewey eye
My Lizzie dear and I.
He must know not, or regret
And will always love his pet.
When on land, I'll have to get
And wink the other eye.

When with loving words he pleads
And persuading intercedes,
Think of those departed deeds
When Lizzie dear and I
But enjoyed a sweet romance.
And if ever we should chance
To meet again in loving glance,
We'll wink the other eye.

PLATONIC FRIENDSHIP

(1915)

Dear girl, though our friendship's platonic,
And all our demeanour is prim,
Our passions are probably chronic,
And, likely, our conscience to dim.
Though eyes look to eyes, with a love's light
That tingles the blood like a tonic,
Though we fondle and coo as the doves might,
Our friendship is purely platonic.

The Idyll of Plato's our border,
Our thoughts are immaculate dreams,
Our caresses devine are in order
To drive out the demon that deems,
To drive out the demons that rule us,
Subduing sensations so chronic,
And show that our feelings but fool us,
Whose friendship is purely platonic.

We'll not be deceived by the tingle
That glows in our veins like a charm,
When our lips and our hands intermingle
There's really no cause for alarm,
'Tis but a true, friendly feeling,
That tempts us with ardour ironic,
Which is really impotent when dealing
With friendship like ours that's platonic.

MATES

A lady remarked that "there was no harm in mates".
In a poor and squalid attic,
Bowed with grief and heavy care,
Sat a weary weeping woman all alone.
She was aged and rheumatic,
Her unkempt and silver hair.
But proclaims a life of trouble nearly done.
"Why so sad and lone and weary?"
"Why alone without a friend?"
"Why remain", I asked, "where callous hunger waits?"
I am Mrs Grundy, dearie!
And my days are at an end,
For married women now believe in mates.

There were times when I was courted,
Both by madam and by miss.
And by widows, wearing weeds, or only grass.
By my hand they were escorted,
To the realms of worldly bliss,
While the blandishments of man were made to pass.
But the times are sadly altered,
Women smoke and swear and bet,
And without my staid assistance tempt their fates.
When I tell them they have faltered,
They reply "You'd better get,
There's surely not the slightest harm in mates".

In the dim and distant ages
When all ladies were demure,
And a chaperon was always in demand,
I have dined with queens and sages
And my place was always sure,
With the lovely leading ladies of the land.
But they now no longer love me,
They have cast me from their lives.
I'm the poor and wretched victim of their hates.
From their homes and hearths they shove me,
For both husbands and their wives,
Now believe there's not the slightest harm in mates.

CHUMS

Friends may often love and leave us,
And, in leaving, love us still.
Some would go, but scorn to grieve us,
Others may remain and will.
There are those endowed of heaven,
Who have filled our hearts with joy.
From whose sacred presence riven,
Make our love and grief alloy.
How the parting pains us,
How our love enchains us,
Joy remains where silent sorrows come.
The only consolation
Is a near relation,
Whom the poets designate as chums.

Father calls his friend a "cronie".
Mother entertains a "mate".
Sister seeks solace only
In a partner at the gate.
Brother has a pal or cobber,

And a Donah or a tart,
Whom he deems a darling robber,
Having commandeered his heart.
But the absent lover
Would a friend discover
One who would relieve the pains that numb.
Sooth with kind caresses,
All our dread and distresses,
And be to her a loving chum.

If in analysing kisses,
One should wish to understand
Where the most delightful bliss is,
Get a chum to take a hand.
Distant friends may peck at noses,
Courtiers at finger tips,
Lovers glow the cheeks with roses
Or linger lovingly on lips.
That's the bliss of blisses,
None like lovers' kisses,
Forceful flight that renders protests dumb.
When they have departed
Why be broken hearted,
When it's so sweet to substitute a chum?

REPLY TO A LETTER FROM MISS EVA WRIGHT

My Dear Miss Evie:-

Though not over-greedy
Your scandal so welcome was short.
I really expected some stories connected
With various phases of court.

You should not dissemble and cause one to tremble
By promising scandal of note
And making one wonder (Excuse the sad blunder)
If you'd really eloped with your pote.

Or if Dobbie (Oh! Never) a swear did
When trumping his partner's ace,
And causing Miss Annie to think him uncanny
And Miss Eva had taken her place.

And then what was better, to make your old debtor
The subject was really too bad,
And call it a scandal, I can't understand all
The reason, if reason you had.

I hope you'll forgive me the rhyming. Believe me
I think you've a penchant for rhymes.
You know you confided a lot more than I did
Concerning poetical times.

You thought of correcting "My dear" but electing
To leave it, as I'm an old spouse was proper.
A wedding might alter the heading
Were your true love the head of the house.

Then "Sir" written plainly might figure more gainly
But I, of the daddy degree
Desire to inherit so honoured a merit
From you as "Dear W.G."

And tell your dear brother when writing another,
If you have no paper to spare,
One page can't command all the choicest of scandal.
No room for a P.S. Now there!

Kind regards to dear Lyton and Arthur the Briton
Mother, Father and Harley B
Your sisters and Bert. Master Bruce and the flirt
From yours truly

W.G.

THE PRODIGAL

When the dark dreary storms of the winter so cold
Have passed from our sphere for a while,
How sweet is the radiant sunshine of gold
That makes the spring morning to smile.

We delight all the more in the music of morn
When the darkness of night is dispersed.
And appreciate better sweet feelings that darn
When we've tasted the better ones first.

When the garden in autumn all flowerless and bare
Is devoid of its beauteous charm,
How we long for its floral adornments to share
When fair spring has extended her arm.

And so 'tis with life, tho' at times we may stray
In the dark mirey winter of vice.
And the springtide of virtue beams out on the way-
How its fondest endearments entice.

And when through the dark nights of sin far away
We are tempted to recklessly roam,
Like the arb of the moring that beacons the day
Is the missive that welcomes us home.

THINK OF ME

(On receiving a lollie with the above motto.)

Like morning in the merry spring
Your loving missive comes,
And sweeter than the birds that sing
The echo of thy name.
But needless was your kind request,
So may it ever be,
For while the east is from the west
I'll ever think of thee.

The days that make thy memory dear
Time's fading power defy.
And in my heart for dye sincere
Thy memory ne'er shall die.
Who was in all the world, my love,
So gladsome then as me.
Say were my heart as hard as steel
Could I but think of thee.

When morning dew is on the flowers,
When evening shadows fall.
When brightly gleam the noontide hours,
When darkness covers all,
My mind on poems and joys of life
Must ever active me.
But still amid this worldly strife
I'll ever think of thee.

APPENDIX III

LETTERS FROM JAMES LAMB "THE PRINCE OF PROSPECTORS" (Ref. CHAPTER 14)

(Transcribed From handwritten letters)

James Lamb
Rosebery Northeast Road
SOUTH AUSTRALIA

18 November 1935

W.G. Manners Esq.
KALGOORLIE

Dear Sir

Seeing your address in *The Western Mail*, I am taking the liberty of penning to you a few lines, to know from you, are you a relation of my dear old friend, a mining engineer of Broken Hill, some 50 or 55 years ago.

He was also at Tarrawingee.

If you are a son of his, kindly write me a few lines to say yes or no.

If you are a son, I'll write to you a long letter.

I have been prospecting since 1852 all over Australia. I am 93 years old last August.

When I get a few lines from you, I'll reply with a long letter. I hope you are. Mrs Manners will remember me.

Hoping to get a letter from you.

Yours truly

JAMES LAMB

18 December 1935

My dear, dear young Mr W.G. Manners (*meaning C.B. Manners*).

I got your letter of date 15 December and I don't ever remember being so pleased to get a letter before, in my life of 94 years of age.

Yes your good dear father was one of the best of good men, I can remember him 50 years past. He was one of the first good mining men to go to the Northern Territory, to report on a lease I had pegged and worked for 16 months.

When I prospected the reef and see gold daily, I went to Bond Springs cattle station 90 miles away. A man named Taylor was the Manager, a man cooking on the station, named Harry Luce and a handyman working on the same station named Jack Thompson.

These two men I engaged to come with me as partners on equal terms of profits of all gold got. I bought 16 bags flour costing £10/10/0 per bag and a bullock.

Now these two men see a drill, for the hole bored in the rock.

Before long we got to work. They soon took to this sort of work and as the stone came up out of the shaft, I dollyed it, then dry-blowed it.

This I was doing for 13 months, not Sundays. Then I got green hides and made them into bags. Now I went to Alice Springs to see the acting warden, a policeman named Bob South. He came back with me to the reef. He see me bag up one ton of ore "seconds". This he reported to the Adelaide Government. Then I gave a teamster £90 - to cart a ton of this ore bagged in green hide bags to Oodnadatta. I left 200 pounds worth of gold with H. Luce and Thompson. Then I started for the city of Adelaide and made it as public as I could, in the newspapers, and amongst brokers.

Now I got word from Mr Hushus, the station master at Oodnadatta. My load of ore has broken down at Bloods Creek. I wrote to Mr Hushus enclosing a sum of money to get it down for me. A few weeks passed away and my ore arrived in Adelaide Railway Station.

I put a bag of this ore in Mr Benda's shop in King William Street, Adelaide and paid him £1 per week, to let the bag of ore lay up against the counter of his shop.

People could take samples and have it assayed for their private information.

Then I sent a bag of two cwt to Broken Hill consigned to Mr W.G. Manners. Then a little later I went to Broken Hill and through Mr Manners and a few of his city friends, it was got into a company.

This caused a big stir in the Hill, then later at a meeting, I proposed that Mr Manners go to MacDonnell Ranges and give a report of this mine and arrange for mining machinery to be sent up according to his instructions.

He got up safely, thousands waiting daily to read his report of this first mine in the Northern Territory. It came thus "gold shining all over the quartz. The means of transit - camels cannot get up so no machinery with camels. It will be good for the prospectors when a railway comes to this country".

Now my dear Mr Manners, tell me, can you find a man in Australia like this man, your loved father W.G. Manners mining engineer. To give a report like this, a true report and from that day, until today, in my old age, love to think of your dear honest father. I love to think of him and your dear good mother. I could fill a newspaper of his good honest deeds in his lifetime.

He has gone, and I am going. I am a stranger and a pilgrim here.

Sometime ago, I got a nice letter from Smiler Hales. He is in England. He has written quite a lot of books and he quotes me "Jim Lamb, the prince of prospectors". He published the first newspaper in Broken Hill, a fine good man.

I wish you and yours a happy Christmas and New Year.

I remain yours truly

JAMES LAMB

24 August 1937

My dear sir, the son of my dear dear friend W. G. Manners of early Broken Hill 56 years ago. I got a letter two years ago from you, I put it away very carefully but could not find it till a month ago.

I would like to pen to you some of the early history of Broken Hill, but I would like to get a few lines of you being still in Kalgoorlie.

I am posting to you a newspaper, the news giving an account of my 95 years of age. Born in Adelaide August 1841 and have prospected all over Australia since 1852. For copper, silver, gold, mica, rubies.

So I will wait till I get a few lines from you to know you are still in business. I am posting the newspaper "The Mail" to you tomorrow.

I remain yours truly

JAMES LAMB

14 September 1937

My dear friend Charles Manners

I am awfully pleased today to be able to pen a few lines to the son of my oldest best friend. In my 95 years of age. Born in Adelaide in 1841 and then in the year 1852, when a boy of 10 years old, with my father John Lamb a builder I worked as a youth with my father. For some few years at the trade, then at 17 years of age I went with an old prospector on the Turon River, in the Snowy Mountains N.S.W.

All the books in this world could not teach this good old man "Jaycob" anything on prospecting, a good true living old gentleman. He one Sunday morning saying "Jimmy you go home to your mother, I go home to my sister in Germany" he placed £400 around my waist, got a passage ticket for me to Port Adelaide. We parted my good dear friend, I got home to Angas Street, in the city Adelaide.

I had no father he had passed away and my dear mother was pleased to see her dear boy again. I stayed with my mother two weeks then went away and prospected all over Australia for 67 years. I came into the Silverton Country 1885 and here I meet Mr Paddy Green, the first man to prospect that country. He was the actual father of this field. I am convinced that no man can more justly claim the title than the late Paddy Green. This man many years ago delved upon the Barrier. Men in their wisdom laughed at him, others more compassionate, but equally incredulous said he had a silver mania engendered by misfortunes upon the great Victorian Goldfields. But Green went on his way unheeding all and firmly convinced, in his own mind that he was on the eve of a great discovery. He knew nothing of chlorides or oxides and sulphides. These to him were an unknown quantity. But he did know galena pure and simple when he saw it. And Thackaringa is wonderfully productive in this precious metal. I believe the first mine worked here was the "Pioneer". Green fossicked out from the surface several hundred tons of good ore from this rightly named claim, but in all his delving he never went 30 ft below the surface. He simply routed out the rich layer or band of silver ore. From this district he sent away several lots of ore, some of which never reached its destination.

Now the rush to the fields when silver was found at Umberumberka. Australia began to grow uneasy, thousands of men looked eagerly for fresh news. Hundreds came in small parties and scattered themselves over the field, but it was not until 1880 that people began to throng to the new Eldorado. Then a steady stream set in, which has never stopped. In 1883 a tremendous rush set in. Terowie, then the terminus of the South Australian Railway Line in this direction was the scene of excitement.

Crowds of eager men thronged the streets in strange apparel. Vehicles of every conceivable description passed daily, even hourly through its busy streets. Hill & Co's coaches were crowded to excess. Teams full of diggers swags pushed on for the great silver fields. Crowds of men placed their "Blueys" upon their backs and tramped off determined to reached the promised land or perish in their efforts.

From Terowie to Silverton was a fearful sight when the winter set in. Teams were bogged all along the track and there were sometimes 20 and 30 at one bad patch commonly called a "Gluepot".

Gluepots were wonderfully plentiful on that track, sometimes as many as 50 horses would be hitched onto a wagon. Whips would be at work, plied by brawny arms. Blasphemy as hot as lava would follow the whips and shovels and spades would be going on in front of the wheels, digging out the team.

Sometimes the wagon would come out and sometimes it would come in sections, torn asunder by the fearful strain.

Then the mail coaches would come creeping along piloted by those princes of the Australian road Nicholas and Wilkinson. Sometimes it would take them nearly a fortnight to get through and it was a living wonder how they ever got along at all.

No man who does not know the back blocks can appreciate the determined pluck and marvellous skill of our mail coach drivers.

A drive on a pitch black night through Stephens Creek with the flood at a banker, a couple of half-broken brutes in the team, and Billy White for a pilot will do more to convince a person than the longest eulogy ever written, on foot, or in the saddle, by coach or by team.

The diggers reached Silverton at last, coming as I did along the old Umberumberka Road and, the dash is made for the tops of the hill, this reached, Silverton lay before us as a mass of canvas stretching in all directions.

From shanties, looking like big sardine tins dotted the whole valley, a mighty throng of restless moving humanity of all nations, all colours and all creeds, bustled about. Some laughing, some sorrowing and many cursing. All busy, and each intent on making money somehow.

Here was a crowd of Irishmen wanting to shout for everybody, or to fight with everybody; there a patient, cute-looking Chinaman or two, soft insinuating, and progressive. Interfering with no one and scoffed at by all. Behold a group of Germans fair faced, half mad with excitement, talking and gesticulating in the most frantic manner as if they had a world of business to do and no time to do it in.

Englishmen were everywhere, strong resolute surly. Determined and arrogant Scotsmen keen, eager and arrogant looking after the Main Chance and dealing in everything. And Australians bold, pushing, boastful, generous and aggressive. Keen as a razor and hardy as Arabs. Jews were there peddling everything under the sun and vowing that they lost money.

My dear friend, a neighbour has now come into my room, with "Smith's Weekly" newspaper, with a full account of me being through the Silverton and Broken Hill country prospecting some 50 years ago.

By me posting to you the newspaper it will be much better than me writing. My age is slowing me down and I cannot write how I could a few years back. Your dear father was a man as a mining engineer, a true good man, to give a true report on a mine. In my good prospecting days I got him to go to the White Range, Tennant Creek, Mt. Bresse, then I wired to some 40 good men prospectors, to meet him and give to him a good reception, and he wrote to me a nice letter and I gave it to Smiler Hales, the editor of the first newspaper in Broken Hill. Hales had him in the newspaper in England as a true honest mining engineer of Australia.

There is no man living that can say a nasty word about W.G. Manners, mining engineer of Broken Hill 50 years ago and the Northern Territory. I am 95 years of age, hoping you and yours, Godspeed.

Yours truly

JAMES LAMB

1937

My dear Charles Manners, son of my dear friend W.G. Manners.

Many thanks for the annual progress report of the Geological Survey for the year 1936. Yes my dear young friend, I prospected in the Musgrave Ranges. The Hartz Ranges. The Petermann Ranges and 80 or 90 miles into West Australia. I was in that country over 50 years ago and on all the camps I left them by cutting my name onto a gum tree and Doctor Bresdow came on one of my camps. He found me, took me to his home and showed me this photo, saying to me "Do you know this photo?". On looking at it I said yes Doctor. How did you get this? He said he took a photo of my camp and asked me how long I was there. I said eleven months. He agreed by saying that I was there a long time as the ashes from my fire would be quite 3 dray loads, if carted away.

I shall try to give you the history of Broken Hill. When I was first in the Hill the population was 700 and when I left the Hill its population was 21,000 picks working on the Hill and down underground. Mr Sleath was president of the miners union which at that time was 13,000 in number.

They took Sleath to Sydney and gave him 2 years in the stockade. Then a Mr Cann was president of the miners union.

Cann became a member of parliament for Broken Hill for years. Mr W.G. Manners and Mr Jack Cann were great friends. Mr W.G. Manners was the first mining engineer on the Broken Hill Line of Lode. So I gave to Mr Jack Cann the first job of work at the Four Mile. I was boss of the Okay and Hidden Treasure Mines. He put down the shaft 200 foot. I paid him £12 per foot. For an Adelaide syndicate I prospected out from Silverton 3 years. Before the Hill was known. Paddy Green was the first prospector in Broken Hill country.

I have had a life of prospecting since the year 1852 all over Australia and I started with my father when 10 years old.

I remain, my dear young friend, yours truly.

JAMES LAMB

P.S.I will try and get a map of the Musgrave country posted to you.

16 March 1938

To Charles Manners, son of my dear old friend W.G. Manners, my dear friend, the finest, the best good man Australia ever had, my dear friend.

I do hope this finds you all well and as happy as this world can make us.

Thanks for your welcome letter and book to hand.

At my age 96 years old, it is nice and good and kind of you to post to me as you have done and do hope you are all well. I have never met a man in my lifetime I've respected like W.G. Manners Esq. mining engineer. I love to think of him as I've known him. I do at times feel sorry he ever had anything to do with that Mr Sim. First a store keeper at Tarrawingee and then a public house in Broken Hill, he was a shake-down, was Mr Sim. But that sort of man is despised by his fellow man for the remainders of his life. Yes I am a helpless old man, hardly able to get about. My first dear wife passed away some 35 years ago at Guildford, West Australia. My eldest son Dave Lamb is a farmer at Pingelly, Western Australia. The second son is in Geraldton, Western Australia. The third son is in Guildford West, living in the old home West Guildford. He is in the Government Railways, an electrical engineer. My daughter lives in Perth, 73 Raglan Road, Mt Lawley. Mrs J. Jury, she has a 10 acre farm six miles out from Perth. Jack Jury is on it. He is a Government perchener, with one son, a wool classer, also a daughter, a lawyer in Perth.

I prospected in and around Silverton 3 years before Broken Hill was found. Then the population was 14, I counted "1" of the fourteen. When I left the Hill, the population in the mines was 40,000 picks working. I'll write you the Hills history. Its first prospector was Paddy Green.

I am going to try and give to you the history of Broken Hill, as your father was the first mining engineer in the Hill 1886-88.

The Hill is 53 years old this year.

It was first prospected by Paddy Green, in the year 1873. The first silver found at Umberumberka. How the diggers reached Silverton -mine managers - one of the wonders of the world - the Broken Hill Proprietary Company. In September 1883 by Charlie Rasp, the town of Wilcannia - the underground workings. The Lead Galleries -Distribution of Wealth - British Broken Hill Proprietary Company - the White Leeds Lime Company, with Mr W.G. Manners engineer. Read "The Broken Hill Lode" written by Smiler Hales, who printed the first newspaper in the Hill, and 38 more mines found. The Terrible Dick Mine, The Rise and Shine Mine, The Okay and Hidden Treasure Mines. Manager Jim Lamb, manager,. The first public house and the second. It will take me six months to write it all down.

JAMES LAMB

APPENDIX IV

White Feather Proprietary Gold Company Ltd Kanowna
How a new company was floated in 1894.

(This material was kindly provided by Mr J.R. MacKenzie of Cottesloe W.A., the son of James MacKenzie of Kanowna. These records were stored down a rabbit warren for "safekeeping" while Mr MacKenzie was away during World War II).

OF THE

White Feather Proprietary Gold Comp any, Ltd.

(TO BE REGISTERED UNDER THE COMPANIES' ACT.)

Situate at White Feather, near Coolgardie, West Australia.

Capital - £12,500 in 50,000 Shares of 5s. each

Of which 25,000 are issued as fully paid-up to the present Proprietors, and 25,000 are offered to the public on the following terms:-

2s. 6d. per Share on application; 2s. 6d. per Share on allotment.

Of the amount subscribed, £5,250 will be placed to the credit of the Company, less the bare flotation expenses, estimated not to exceed £400.

Provisional Directors:

HON. WM. LAIDLEY, M.L.C.
ALFRED HARCOURT, ESQ.
ROBERT AMOS, ESQ.

W. E. WALMSLEY, ESQ. (MOREE)
R. G. WATKINS, ESQ.
J. B. CHRISTIAN, ESQ.

Bankers: BANK OF NEW SOUTH WALES.

Brokers:

SYDNEY—FELIX RANDLE, Union Bank Chambers, 68½ Pitt-st.; MELBOURNE—ALFRED TOLHURST, Prell's Buildings, 58 Queen-st.; ADELAIDE - WILKINSON & HARRISON, G. M. ... us ... Nolan ... in ...

Secretary (*pro. tem.*): LEONARD DODDS, 131 Pitt-street, Sydney.

THIS Company is being formed to acquire from the present holders their entire interest in Gold Mining Lease No. 533 (known as McKenzie's, and having an area of 15 acres), and immediately adjoining McAuliffe's Lease No. 156, which was recently purchased by Lord Percy Douglas and Mr. W. A. Mercer for a London Syndicate. The purchase money paid for this property being £6500 in cash. The other adjoining blocks have been purchased at large prices by Adelaide Companies. From advices since received, great disappointment was felt by English representatives in not having acquired this property.

As will have been seen, by the reports appearing in the daily press, the White Feather portion of the Coolgardie goldfield has been brought into great prominence of late, which has been largely increased by the series of purchases made on behalf of English capitalists.

From the subjoined Reports by well-known mining authorities, there are two known reefs running through the property, varying in size from 1 to 2 feet, in a N.W. and S.E. course. There also appears to be a third vertical reef, but upon which no work has been done up to the present.

The property also contains some very rich leaders, and doubtless some of these will be found to junction in depth, giving place to another rich reef. It will be noticed that the original proprietor has made the property pay from the start by merely dollying the stone in a primitive manner. The adjoining properties have also been paying in a similar way.

There is an ample supply of Mining and Milling timber in the immediate vicinity. As to water, a most important factor in this district, there is a good and permanent supply obtainable from a lake about three miles from the mine. Besides which, it is confidently expected that a sufficient supply will be met with in sinking at a moderate depth.

It is estimated that the Capital to be placed to the credit of the Company, **£5250**, will be more than ample to open up the mine, erect the necessary machinery, and place it in a dividend-paying position, when the smallness of the capital, **£12,500**, will tell most favourably in the results.

The Company will be declared formed when 20,000 shares have been applied for.

After allotment of the shares, a meeting of allottees will be convened by the Provisional Directors for the purpose of electing Directors and adopting Memorandum and Articles of Association.

The only Agreement for the sale of the property is one dated 12th May, 1894, between J. R. M'Kenzie and others, which can be inspected at the Company's Office, 131 Pitt-street, Sydney.

Application for Shares to be made on the accompanying form to any of the above-named brokers.

SYDNEY, June, 1894.

Form of Application for Shares.

TO THE PROVISIONAL DIRECTORS OF THE

WHITE FEATHER PROPRIETARY GOLD COMPANY, LIMITED.

(To be registered under the Companies' Act.)

GENTLEMEN,—Herewith I hand you the sum of £........................ being a deposit of the application fee of 2s. 6d. per share on........................ shares in the above-named intended Company. I request you to allot me that number of shares upon the terms of the prospectus of such intended Company, to which I agree to be bound, and I hereby agree to accept the same or any smaller number that may be allotted to me, and to pay the balance of 2s. 6d. per share on allotment, as provided by the said prospectus; and I authorise you to register me as holder of the said shares on the registration of the Company, and I agree to be bound by the Memorandum and Articles of Association of the Company, which shall be agreed to in terms of the prospectus.

Dated this........................day of........................, 1894.

Name in full........................

Address........................

Description........................

Signature........................

P R S

White Feather Propriе Company, Limit

(To be under ... act),

SITUATE AT WHITE FEA NEAR COOLGARDIE, WEST AUSTRALIA.

Capital - - - - £12,500, in 50,000 Shares of 5s. each.

Of which 25,000 are issued as fully paid up to the present proprietors, and 25,000 are offered to the public on the foll wing terms :—2s. 6d. per Share on application, 2s. 6d. per Share on allotment. Of the amount subscribed, £5250 will be placed to the credit of the Company, less the bare flotation expenses, estimated not to exceed £400.

PROVISIONAL DIRECTORS:

HON. WM. LAIDLEY, M.L.C.	W. E. WALMSLEY, Esq. (Norre).
ALFRED HARCOURT, Esq.	R. G. WATKINS, Esq.
ROBERT AMOS, Esq.	J. P. CHRISTIAN, Esq.

BANKERS:
BANK OF NEW SOUTH WALES.

BROKERS:
SYDNEY: FELIX RANDLE, Union Bank Chambers, 68½ Pitt Street.
MELBOURNE: ALFRED TOLHURST, Prell's Buildings, 58 Queen Street.
ADELAIDE: WILKINSON AND HARRISON, Exchange.
BALLARAT: J. D. WOOLCOTT, Exchange.
BENDIGO: D. DWYER, Beehive Exchange.

SECRETARY (pro tem.) LEONARD DODDS, 131 Pitt Street, Sydney.

THIS Company is being formed to acquire from the present holders their entire interest in Gold Mining lease No. 533 (known as M'Kenzie's, and having an area of 15 acres), and immediately adjoining M'Auliffe's Lease No. 156, which was recently purchased by Lord Percy Douglas and Mr. W. A. Mercer for a London Syndicate. The purchase money paid for this property bein £6500 in cash. The other adjoining blocks have been purchased at large prices by Adelaide Compan es. From advices since received great disappointment was felt by English representatives in not having acquired this property.

As will have been seen by the reports appearing in the daily press, the White Feather portion of the Coolgardie goldfield has been brought into great prominence of late, which has been largely increased by the series of purchases made on behalf of English capitalists.

From the subjoined Reports by well-known mining authorities, there are two known reefs running through the property, varying in size from 1 to 2 feet, in a N.W. and S.E. course. There also appears to be a third vertical reef, but upon which no work has been done up to the present.

The property also contains some very rich leaders, and doubtless some of these will be found to junction in depth, giving place to another rich reef. It will be noticed that the original proprietor has made the property pay from the start by merely dollying the stone in a primitive manner. The adjoining properties have also been paying in a similar way.

There is an ample supply of Mining and Milling Timber in the immediate vicinity. As to water, a most important factor in this district, there is a good and permanent supply obtainable from a lake about three miles from the mine. Besides which, it is confidently expected that a sufficient supply will be met with in sinking at a moderate depth.

It is estimated that the Capital to be placed to the credit of the Company, **£5250**, will be more than ample to open up the mine, erect the necessary machinery, and place it in a dividend-paying position, when the smallness of the capital, **£12,500**, will tell most favourably in the results.

The Company will be declared formed when 20,000 shares have been applied for.

After allotment of the shares, a meeting of the allottees will be convened ... Directors for the purpose of electing Directors and adopting Memorandum and Arti ... ation.

The only agreement for the sale of the property is one dated 12th May ... J. R. McKenzie and others, which can be inspected at the Company's Office, 131 Pitt S...

Application for shares to be made on the accompanying form to any of the abo ... Brokers.

SYDNEY, June, 1894.

Report by W. H. Corbould, M.A.I.M.E.

May 9, 1894.

LEONARD DODDS, Esq., Sydney,

DEAR SIR,—I beg to submit my report on M'Kenzie's Claim, M.L., 533. The property is situated at the White Feather, 35 miles N.E. of Coolgardie. The property has an area of 15 acres. The lodes being worked at present are "Blanket lodes," carrying very rich gold. The country is quartzite, and, in places, the gold is showing in the quartzite. The strike of the lodes is about N.W. and S.E., dipping to the N.E. There appears to be a vertical lode on the property, but no work has been done on it, so that I cannot say with certainty. Since taking up the property, M'Kenzie has made a few hundred pounds profit, after paying men's wages. The two adjoining claims—Doyle's and M'Auliffe's—have also paid all expenses and paid good dividends to the owners. The work done on M'Kenzie's is very primitive, and the Company will have to start and work the mine in a systematic way, which will mean the outlay of a thousand or two pounds. Rich stone is showing in the face of the prospectors' workings, and it may be possible to dolly enough gold to pay further prospecting of the mine at a deeper level. In conclusion, the property is a good mining venture, worthy of being tested by a Company.

Yours faithfully,
(Signed) *W. H. CORBOULD.*

Report by Robert Pettigrew,

MINE MANAGER OF THE MARITANA AND NAPIER GOLD MINING COMPANY.

DEAR SIRS,—The mine that I visited a short time ago is situated at the White Feather, 35 miles north of Coolgardie. This property is held by Mr. M'Kenzie as ...ing lease, No. 533, and consists of 15 acres. It is bounded on the S.E. by M'Auliffe's property ...h has lately been purchased by Mr. W. A. Mercer, of Coolgardie (of the firm of Bewick, Moreing and Co., London), and it is bounded on the S.W. by Doyle's property (which was lately bought by an Adelaide Co.) it is bounded on the N.W. by Meiklejohn's property, which has been purchased by Messrs. Braidwood and Bedford, of Adelaide. All three properties are raising good stone, and have dollied out sufficient gold to pay all working expenses for the last five months. I consider lease No. 533 is equal in value to any of the above blocks mentioned. There are two known reefs running through this property, and a number of rich veins. The reefs are from 12 in. to 2 feet wide.

Very little work was done on those reefs at the time of my first visit, but on my second visit a large amount of trenching had been done, exposing the reefs. The stone taken from these trenches is very r... in gold; this gold is well disseminated throughout the stone, and, in my opinion, is rich enough for dollying. Mr. M'Kenzie has been paying all working expenses by this means.

The strike of the reefs is west by north, with an underlay of 20 min. nor'-east. The hanging wall of the reef is quartzite, the foot wall is diorite. Samples taken from the hanging wall prove this quartzite to be gold bearing, and in Doyle's block, 30 feet from the boundary line of McKenzie's, I saw large blocks of this quartzite broken from the hanging wall which was sufficiently ... for dollying.

There is a good supply of mining and milling timber on and in the vicinit... property to last for many years.

The water at the White Feather is obtained from a lake three miles away, w... supply is always obtainable, but I have no hesitation in saying that water in sufficient qu... be got in sinking at a depth of 150 or 200 feet.

Speaking of the White Feather generally, as a reefing field, the nature of th... to believe the gold will not be found in patches, but uniformly distributed throughout ... also think the reefs will continue to a great depth, and will increase in width and ... is attained. Developments up to the date of my last visit confirm the above opinion, and all mi... men of any standing who have visited this field are very favourably impressed with it as a re... field.

There is no doubt the leaders on this and the adjoining properties are immensely rich, and all leading towards the main reefs.

GENTLEMEN,—I have no hesitation in recommending this property to investors as a valuable mining property.

Yours respectfully,

HANNAN'S FIND, May 9th, 1894. (Signed) *Robert Pettigrew,*
Mining Manager Maritana and Napier Gold Mining Company.

Report by J. Treloar,

MANAGER OF THE AGNES GOLD MINING COMPANY,
HANNAN'S.

I have visited this property, and have great pleasure in reporting as ...

The mine is situated about 16 miles from Hannan's Rush, and the pl... Feather. The lease, No. 533, adjoins on the S.E. McAuliffe's lease, which has ... Lord Percy Douglas and Mr. Mercer for an English Company, and on the S.W. ... just been floated in Adelaide.

The property consists of 15 acres, and there has been large quantities of alluvial workings on the course of those reefs. Very little work has been done or stone has been taken out for dollying. There are two reefs on the property, varyi... to 2 feet, widening out as they go down, and can be traced right through the prop...

The strike of the lode is N.W. by S.E., with an underlay to the N.E. ... that some of those will junction at a depth, and more than probable a rich reef ...

The gold is well distributed through the stone, and is a good battery ... gold, and should yield several ounces of gold to the ton of stone.

I saw the lessee taking a quantity of gold to the bank that had been dollied ... was from small gold up to ½oz. pieces. I believe this property to be one of the best situated, for most of the rich leaders running through the properties on the White Feather. Therefore, I am quite certain the property is a very valuable one, and I can strongly recommend it to the investing public.

I may mention that M'Kenzie has made it pay from the start.

(Signed) J. TRELOAR,
Manager of the Agnes Gold Mining Company.

SYDNEY, 22ND JUNE, 1894.

THE PROVISIONAL DIRECTORS,
WHITE FEATHER PROPRIETARY GOLD COMPANY, LIMITED,
SYDNEY.

DEAR SIRS,—

The following is my opinion respecting the reefs in the "White Feather" district, near Coolgardie, Western Australia :—

Regarding their permanence—that is assured. Timber for fuel and mining purposes, without question, the best on the "Yilgarn" Goldfield, and a lasting supply of water within a few miles.

Although many of the lodes carry good dollying stone, their character is not patchy, but good general milling quartz, the gold being both coarse and fine and equally distributed through the vein. Body of crushing stuff varies in thickness from 2 to 4 feet, and, as a rule, increases in size with depth, with no falling off in gold yield.

These reefs should give excellent results, having the undeniable advantages of timber and water.

I am, Yours faithfully, A. GOLDIE.

FORM OF APPLICATION FOR SHARES.

TO THE PROVISIONAL DIRECTORS OF THE

WHITE FEATHER PROPRIETARY GOLD COMPANY, LIMITED.

(To be Registered under the Companies' Act.)

GENTLEMEN,—Herewith I hand you the sum of £......................., being a deposit of the application fee of 2s. 6d. per share on .... shares in the above-named intended Company. I request you to allot me that number of shares upon the terms of the prospectus of such intended Company, to which I agree to be bound, and I hereby agree to accept the same or any smaller number that may be allotted to me and to pay the balance of 2s. 6d. per share on allotment, as provided by the said prospectus; and I authorise you to register me as holder of the said shares on the registration of the Company, and I agree to be bound by the Memorandum and Articles of Association of the Company, which shall be agreed to in terms of the prospectus.

Dated this.......................day of.......................1894.

Name in full.......................

Address.......................

Description.......................

Signature.......................

Dear BULLETIN,—The charge ... was a Maoriland squatter. He ...

Amusements ... Miscellaneous.

THE WEST AUS'

McKenzie's Gold Mines, Limited.

Registered by C. T. Stiskings, 1, Copthall-buildings, E.C., with a capital of £75,000 in £1 shares. Object, to adopt and carry into effect an agreement, made February 19, 1895, between the Western Australian Development Corporation, Limited, of the one part, and H. E. Walker, on behalf of this company, of the other part, for the acquisition of certain gold mining rights and leases, claims, and property in Western Australia, and to carry on business as miners, and smelters, and the making marketable of minerals, metals, and metallic substances of every description; to acquire and turn to account inventions; as company promoters, stock and share dealers, &c. The first signatories are:—W. M. N. Young, Bart., 45, Matheson-road, Kensington, W.; W. E. M. Farrar, 6, Moorgate-street, E.C.; J. H. Ivery, M. and C.E., Guildford; H. E. Walker, 20, Victoria street, Westminster; J. R. Kingstead, Upton-park, Slough; W. J. Lane, 68, Dynevor-road, Stoke Newington; H. D. Hepburn, 7 and 8, Pancras-lane, E.C. The first directors—to be not less than two nor more than seven—are to be nominated by the signatories. Qualification, 100 shares. Remuneration, £100 each per annum—chairman, £150—with a percentage of the profits, divisible.

East Nigel Gold Mining Company, Ltd.

Registered February 18, by Foster, Grave ... 15, Finsbury Circus, E.C., with a ...

...moothly and well. The ... and boiler to be used for the bat... have arrived on the mine, and are now ready for erecting.

Messrs. MacKenzie and Clark have just made an extremely rich discovery on a block about ¾ of a mile north of McAuliffe's, at the White Feather. The block consists of twelve acres, and during the last few weeks prospecting operations have been vigorously proceeding upon it. The reef is two feet in width, and carries good gold throughout, while at a depth of nine feet magnificent stone has been struck. Mr. MacKenzie arrived in Coolgardie last Tuesday with some ... specimens from the new find. Some idea of the value of the stone may be gathered from the fact that six ounces of gold was chipped from 30lbs. weight, without any crushing. Pieces nearly an ounce in weight were also broken from the stone. Right through the reef averages ten ounces to the ton, and altogether the discovery is regarded as one of the richest yet made in the district.

Mr. J. R. MacKenzie has just completed for a Perth gentleman, the purchase of Robinson and Collier's claim at the Ninety-Mile, for the sum of £5000 and a tenth interest.

Macdonald

... Tatham, Oblein and Nash, Queen ...

McKENZIE GOLD MINES, Ltd. (43,341).

This Company was registered on February 20th, with a capital of £75,000, in £1 shares, to enter into an agreement with the Western Australian Development Corporation, Ltd., and to acquire, work, develop, and turn to account a certain gold-minin[g] lease, mining rights, claims, and property in Western Austra[lia]. The subscribers are:—

	Shares.
W. M. N. Young, Bart., 45, Matheson Road, W. ...	... 1
F. M. Farrar, 6, Moorgate Street, E.C., gentleman	... 1
J. H. Ivory, M. & C. E., Guildford	... 1
H. E. Walker, 20, Victoria Street, S.W., secretary	... 1
J. R. Kingstead, Upton Park, Slough, gentleman ...	... 1
W. J. Lane, 68, Dynevor Road, N., cashier ...	... 1
H. F. Hepburn, 7 & 8, Pancras Lane, E.C., stationer	... 1

The number of directors is not to be less than two nor more than seven; the subscribers are to appoint the first; qualification, 100 shares; remuneration, £100 each per annum, and £50 extra for the chairman. Registered by Wilson, Bristows and Carpmael, 1, Copthall Buildings, E.C.

I the undersigned Manager of the Robinsons Gold Mines hereby agree to pay Jas R. Mackenzie the sum of Fifty pounds Stg in payment of Gold Mining lease No 820 situated at the White Feather the above amount to be paid on the ~~completion of the~~ payment of first dividend ~~first crushing~~ on the above Companies leases known as the Cocktail & Busker dated 4th June 1895

N W Harker

№ 19101 WESTERN AUSTRALIA.

Fee—One Pound.

West Australian Goldfield.

MINER'S RIGHT.

Issued to J. R. Mackenzie

under the provisions of "The Goldfields Act, 1886;" to be in force until 3 Sept 1896

Date of issue. 4 Sept 1895

[Not Transferable.]

J Campbell [illegible]

Kanowna.
8 April 1897.

Dear McKenzie

I recd your note of 30th inst all right for which thanks.

<u>Re Main Reefs</u>

I have just bought 1000 @ 6/6, and consider there ought to be a bit of a rise in them shortly, the water is increasing & now I am able to run 2 shifts.

The Robinson battery is running steadily, and so far has turned out 2864 oz of smelted gold in 5 crushings.

The Feather is very dead & seems unable to liven up again.

I suppose you have heard that I am contesting this Electorate.

I shall be very glad of your support & vote, and I think that you will give of assistance to me in this matter, as I am well identified with the District.

Kind regards from
Yours truly

C. H. W. Harper

The Robinson declared a divd of 6d per share. Your old claim is going for 3 months.

PALACE HOTEL KALGOORLIE

Kalgoorlie. W.A.

August 15th 1899

Dear Jim

your letter of the 11th inst to hand In the event of our getting the tribute I suppose it will be necessary to put on men to comply with labour Conditions at once I should think three months, even that would give very little time, do your best for the option of extention It will be necessary for you to forward transfer of land ~~to deed~~ or Bank before the money will be paid over to you or me. the balance of half share in Hannans Land Coy you can pay into N S Wales Bank Perth New Castle St Land deeds you can place also at the same Bank Jim the Palace here requires painting etc similar to the Esplanade would you kindly furnish me with particulars of how the latter place is being done price & or get the Contractor to do so it might lead to some work for him I would like particulars about the paper as I think it would suit here well I might ask Brown to go round & see it what did that floor felting cost you per yard

P.S. A large number were at the races on Saturday & the racing excellent - fancy 3 a dead heat your friends luck was generally bad What about the share Market dont it make one wild to stand bye & look on

What a dreadful botchery Cavanagh made of the drain pipes to cost so much also plastering I am sorry the Bar trade was so bad on Friday I hope Saturday made up for it I think you ought to come up and see the show started if we get the tribute

I had a letter from Jack they say that the lathe is sold but will not divulg the buyers name I wonder if it is myself they mean? ask Jack to send the Bill for addvertisment along to the Coy & I am sorry he did not succeed As you were present and I do not like writing to Hartley the Engineer would you ask him to send the lathe along at the price agreed on viz £100— A Consign, for W F M Reef Care Long & Co Kanowna Hartley would be better to Consign in some one else name I would not like ~~them~~ agents to find out that they were done out of the Com, but it serves them right they loaded it so much I think you were lucky to sell the Land Kanowna is dreadfully quiet & very little hope of immediate improvement Kind regards to Mr & Mrs Folk yours truly

C N W Harper

Kanowna
28/8/99

Dear Jim

Returned home all right. Warden Davis from Broad Arrow is acting here so the exemption A may be heard Tuesday or Wednesday. Charley Osmetti one of the injured men has taken a change for the worse & is now in a critical condition.

Re Scrip Hannans Land Coy as there is only one Director here to sign transfer you had better keep the (1000) and give me an acknowledgement for for same and place the five hundred to my Credit or rather safe custody at the U S W Bank you might also place deeds New Castle street land there. The Call of £200 has been made & paid by some of them for the new building will you pay yours in there with the balance owing to me & I shall pay the full amount of Call here. Re the additions as per plan to Hotel I will see you in a few days up here when we can discuss the matter.

Re my account get Mr Holt to send it along when I shall forward cheque for same. Kind regards to all.
I think it would be worth your while to run up. Dunney ...

This is to certify that we the undersigned are partners in the working on Tribute of leases 3X 60X 46X and 81X on the East Coolgardie Goldfield as per agreement dated Septr 1899 between Jas. R. Mackenzie and Henry J. Saunders attorney for W.A. Trust Coy and we agree to equally divide all profits or equally share all losses that may be incurred in the working of the said leases —
dated at Kanowna the 27th Oct 1899

Jas. R. Mackenzie

N W Harper

Hotel Esplanade,

Bazaar Terrace,

Perth, W.A.

Feby 18th 1900

Received from Jas. R. Mackenzie scrip for One thousand Shares No. 18001 to 19000 in the Harmans Land Coy Ltd. Scrip for Fifteen Hundred (1500) Shares in new Coy. to be returned to Jas. R. Mackenzie on demand

N W Harper

APPENDIX V

Summary Pages of The 4cm Thick 1897 Findings Regarding Alluvial Workings on Mining Leases

ROYAL COMMISSION ON MINING.

To His Excellency Lieut.-Colonel Sir Gerard Smith, Knight Commander of the Most Distinguished Order of Saint Michael and Saint George, Governor and Commander-in-Chief in and over the Colony of Western Australia and its Dependencies, &c., &c., &c.

MAY IT PLEASE YOUR EXCELLENCY,

In accordance with the terms of Letters Patent dated the 18th August, 1897, empowering us to inquire into and report as to the best mode in which assistance can be rendered to develop the auriferous resources of this Colony of Western Australia, we, the undersigned, have the honour to submit the following Report:—

1. The preliminary meeting took place in the Committee Room of the Legislative Assembly on the 1st September, 1897, and since that date we have held 135 sittings of the full Commission, in addition to numerous Sub-Committee Meetings.

2. On the 10th November, 1897, the Hon. E. H. Wittenoom, M.L.C., resigned his position as President, and Mr. C. H. Rason, M.L.A., was appointed President in his stead. Having to attend the sittings of the Federal Council in Sydney, Mr. A. H. Henning resigned his seat upon the Commission prior to the first sitting, and Mr R. H. Barrett was appointed to fill the vacancy. Owing to pressure of private business Mr. R. P. Young, on the 6th February, 1898, resigned his seat on the Commission, and was succeeded by Mr. W. H. Gale.

3. After carefully going through the Goldfields Act, 1895, and Regulations, and deciding upon the various Sections of the Act and Clauses of the Regulations which, in our opinion, would require amendment, we commenced to take evidence in Perth, and likewise distributed throughout the Goldfields a list of printed questions, with the view to elicit written evidence on a variety of subjects bearing on our investigations from those who were not in a position to attend personally before us. After sitting in Perth for some weeks it was deemed desirable to proceed to the various Goldfields Centres and take evidence on the spot. Accordingly we visited the Yilgarn, the Eastern, and the Central Goldfields, and took evidence at Southern Cross, Coolgardie, Kalgoorlie, Cue, Mt. Magnet and district. Unfortunately, the time at our disposal would not allow us to visit any other centres, and we were reluctantly compelled to abandon our contemplated visits to Menzies, Lawlers, the Norseman, Dundas, Yalgoo, Northampton, the Collie Coalfield, and other places.

4. We would like to point out, before proceeding further, that the time placed at our disposal was quite inadequate to enable full justice being done to the large field of subjects embraced by the terms of our Commission. The last Victorian Royal Commission, which was appointed to perform exactly similar work to that remitted to us, occupied considerably over two years in its investigations. Nevertheless, although we have been unable to go as exhaustively into the various questions bound up with the

welfare of the mining industry as we could have wished, we trust that the recommendations submitted herewith will, if carried into effect, materially assist the development of the vast auriferous resources of this Colony.

5. The evidence submitted is of necessity voluminous, but the greatest care has been taken to select as witnesses the most experienced and reliable mining men in the different localities visited. Before visiting any field we communicated with the various Associations connected with mining, and invited them to send representative men as witnesses. In addition to the individual views of these gentlemen, we obtained, in writing, the opinions of the various Associations in their corporeal capacity. The evidence which we append should therefore prove of great value to those whose duty it will be to take part in amending the Mining Laws of the Colony.

6. We now beg to submit to Your Excellency the result of our inquiries, which we have classified under various heads, together with the recommendations made in connection therewith.

THE ALLUVIAL QUESTION.

7. When prosecuting our inquiries on the Goldfields we paid particular attention to the working of Section 36 of the Goldfields Act, 1895, which gives to a miner the right, subject to the Regulations, of entering upon a lease to within 50 feet of any reef or lode situate thereon for the purpose of searching for and obtaining alluvial gold. The evidence adduced varied. Whilst mine managers and mine owners, as a body, thought that the leaseholder should be entitled to all gold deposits found within the perpendicular of his pegs, evidence was given individually by some mine managers to show that no harm had resulted in the past through the holders of Miners' Rights working the surface alluvial on leases. The evidence of all the representatives of the working miners, on the contrary, was unanimously in favour of the right conferred by the Act, to enter on any lease to within fifty feet of the reef or lode, not being interfered with. Taken as a whole the evidence was to the effect that very little, if any, friction had been caused in the past by the provisions of Section 36, but it was pointed out that there was certain to be trouble in the future in the event of alluvial gold being found on leases in quantity. We quote the evidence of one representative Mine Manager on this subject:—

> "8517. Do you approve of the present system of giving a dual title to the same piece of ground whereby alluvial diggers may enter on existing leases, and new leases for lode mining may be granted over existing alluvial claims?—I do not. There is going to be a conflict if there has not been a conflict already. If there is no gold they may work amicably together, but if there is gold trouble will certainly arise."

> "8550. * * * * In your answer to question 25 you apparently fear some conflict between leaseholders and the alluvial diggers if the present state of giving a dual title to the same piece of ground is continued. We have evidence to show that so far there has been no conflict. Do you think we may fairly judge of the future by the experience of the past?—I know that you must be a more peace-loving people here than anywhere else. In my experience, wherever gold is found a conflict ensues. I am exceedingly surprised that your experience is not the same."

8. It was after we had concluded taking evidence on the fields that the trouble between the alluvial miners and the leaseholders, which has become historical, occurred in Kalgoorlie and district. There is little doubt that while alluvial mining was confined to surface dryblowing, the mine owners, as a body, experienced no inconvenience by the presence of alluvial diggers working on their leases. Indeed, it came to our

knowledge that various managers of leases taken up under a former Act, which gave the holder the exclusive right to all the gold within the four pegs of the lease, were not only permitting miners to work the surface alluvial within certain limits, but were even encouraging them to do so.

9. The finds of deep alluvial at Kanowna and Bulong, and on the Ivanhoe Venture Lease at Kalgoorlie, have, however, put a different aspect on the question, and we are of opinion that if the present system of dual titles be perpetuated, it will really end disastrously to the mining industry by bringing about a feeling of insecurity of title, which must prove fatal to the best interests of all concerned.

10. Having given careful consideration to this very difficult subject, and having weighed well both sides of the question, we have come to the conclusion that the granting of dual titles is derogatory to the successful development of the mining industry, and should be prevented in the future. We therefore recommend, subject to the conditions which follow, that Section 36 should be repealed, and a provision substituted allowing the holders of leases, prospecting reward claims, and ordinary claims, the right to all gold deposits within the perpendicular of their pegs. We have refrained from making any recommendations with regard to the present dispute between the leaseholders and the alluvial miners in connection with the leases issued under the 1895 Act, as we consider it a question which should be left to the decision of the Supreme Court and the wisdom of Parliament.

11. Having given our opinion that the system of dual titles should be abolished, the next question to consider is how to safe-guard the rights of the alluvial diggers so as to ensure that in the future alluvial ground will not be taken up in leases, but will be kept open for the benefit of the alluvial miner. Everyone acknowledges that the greatest facilities must be given to the alluvial men to enable them to pursue their calling, and that it is to the best interests of the country to encourage the settlement of this most desirable class of colonists in our midst. It is also generally allowed that the present trouble between lessees and alluvial miners has occurred mostly through the mistake made in leasing lands almost exclusively alluvial, which ought never to have been leased.

12. In order to obviate this in the future, and to encourage alluvial mining to the utmost, we are of opinion that no leases should be granted over exclusively alluvial ground, and that before any application for a lease be approved of, care should be taken to ascertain by inspection whether or not the ground applied for is known to contain, or is likely to develop alluvial gold.

13. We further recommend (*a*) that no lease be granted on the present existing Goldfields where alluvial is known to exist for a period of twelve months from the passing of the proposed Amending Act, and (*b*) that no lease be issued on any Goldfield which may be discovered hereafter for twelve months from the date of the proclamation of such field.

14. In a subsequent portion of this Report we have recommended the appointment of local Mining Boards, and we would now suggest that these Boards should be empowered to advise the Department as to what ground should be reserved for alluvial mining, and what areas should be thrown open for leasing. The members of such Boards, from their local knowledge, would, we are sure, be of great assistance to the Department in an advisory capacity.

15. We have given the vexed question of the definition of "Alluvial" careful consideration, and we now recommend in connection therewith that all gold, except such as is found in a seam, lode, dike, or quartz reef or vein, shall be considered alluvial gold.

APPENDIX VI

CONTENTS PAGE

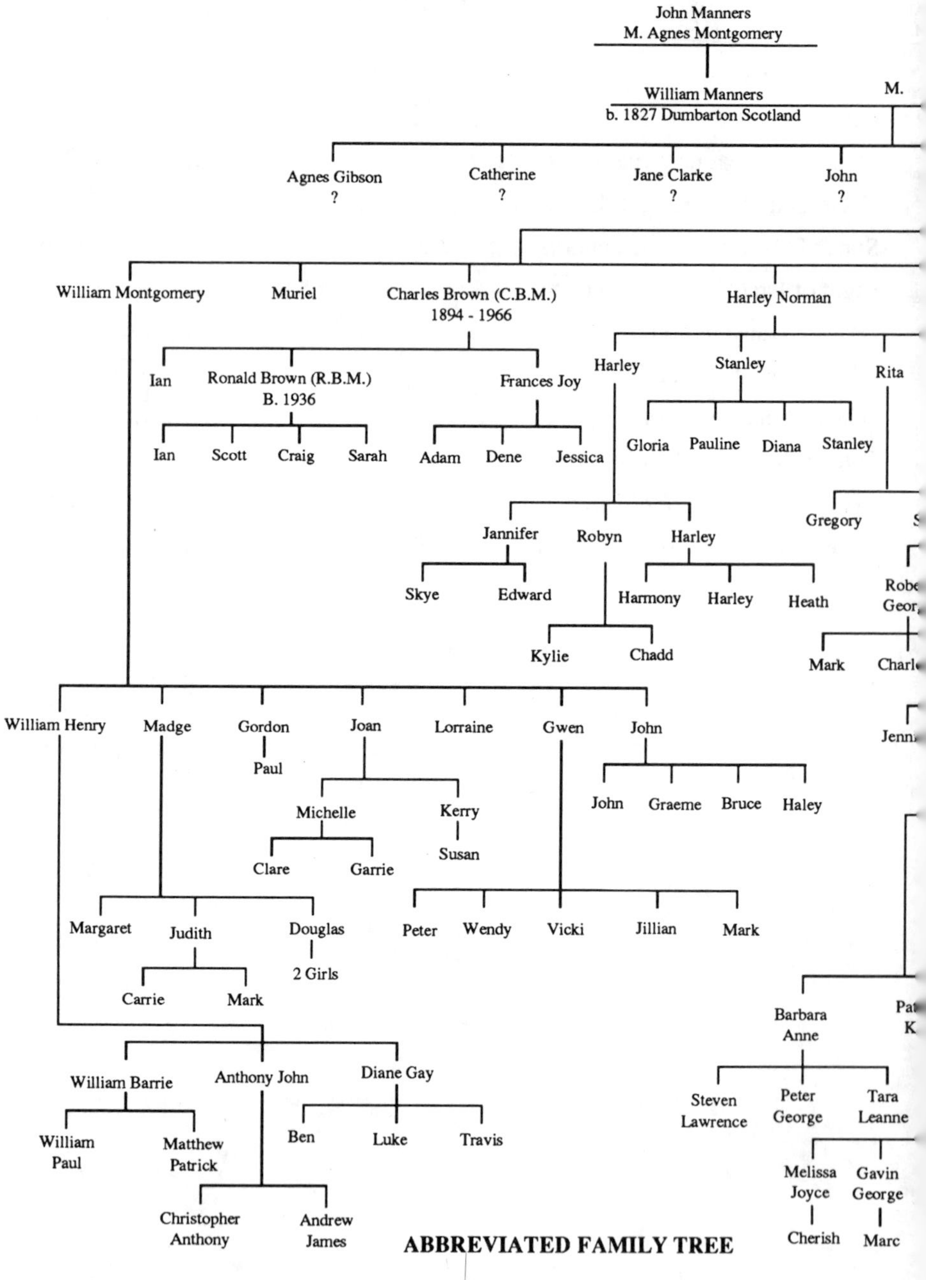

ABBREVIATED FAMILY TREE

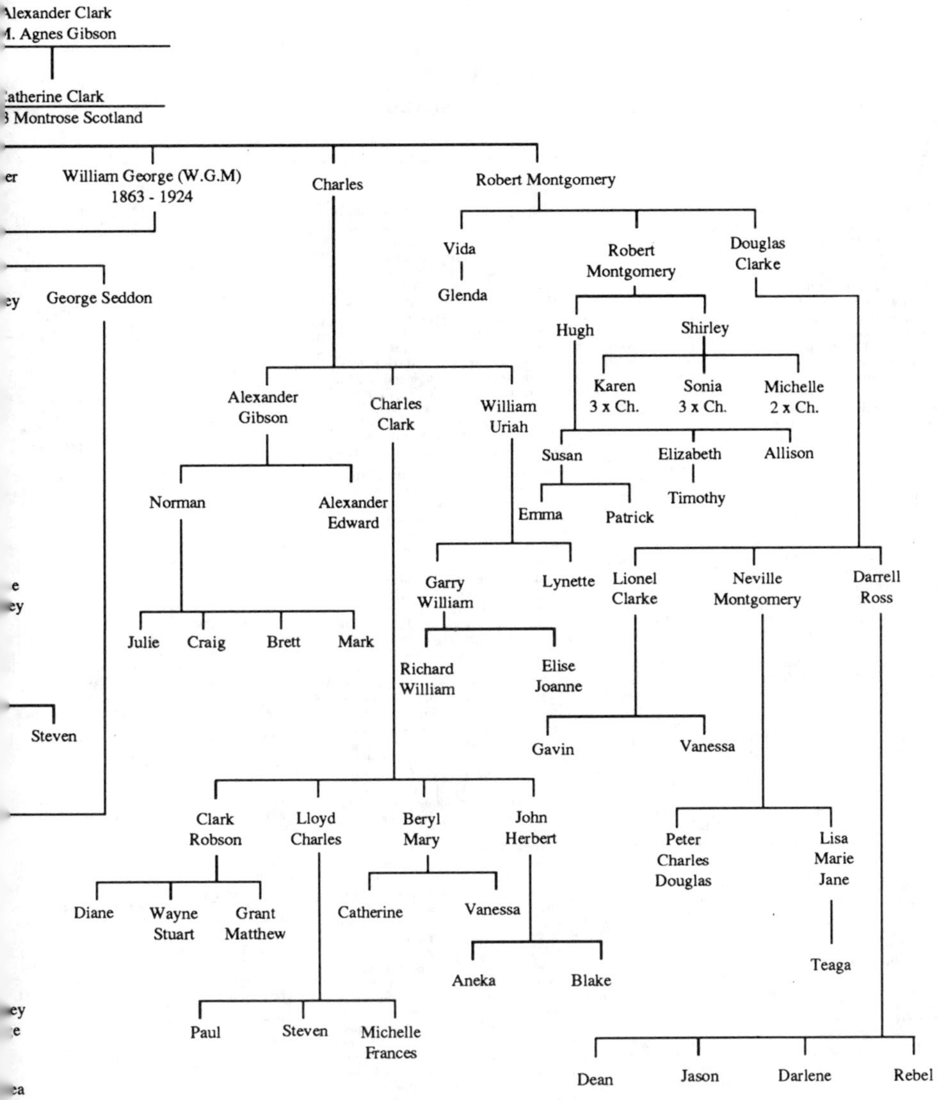

(Space limitations deleted marriage partner's names, and the repetition of "Manners" throughout. Either "Manners" or new family name needs to be added to the christian names listed.)

THE ST. GEORGE TREATMENT PLANT (W.A.).

By Percy Ifould.

One of the latest treatment plants which has been installed on the West Australian goldfields is that recently erected by the Boulder No. 1 Co. on their St. George mine at Mt. Magnet. As this plant is designed from the latest goldfields practice, and differs in some respects from the usual type of mill operating on the northern fields, a brief description may perhaps be of interest to the readers of the "Mining Standard."

Briefly, the treatment is as follows:—The ore, which is practically all mined from the oxidised zone, is coarse-crushed by a breaker, milled in a stamp battery through coarse screens, the pulp passing over amalgamated copper plates into Wheeler grinding pans, where is is all slimed, and the slime product unwatered, agitated with cyanide solution, and filter-pressed in a vacuum type of press. The gold solution is precipitated in ordinary zinc extractor boxes.

Ore-breaking and Transport.—The poppet heads are of the gallows frame type, built of jarrah, and are 60ft. high to the pit head sheaves, with a brace 19ft. above the collar of the shaft. The ore is dumped by means of self-dumping skips on to a grizzley 12ft. long and 7ft. wide, set at an angle of 40deg. to the horizontal. The grizzley is fitted with taper bars, having a 1¼in. space. The oversize passes on to a feed floor, built flush with the top of the crusher jaw, and lined with ¼in. steel plate. The rock breaker is a 12in. x 8in. Blake pattern jaw crusher, running at a speed of 250 r.p.m. The crushed ore, together with the undersize from the grizzley, gravitates on to an 18in. belt conveyor, 25ft. long, having a rise of 6ft. This delivers the ore into the hopper of a Krupp pattern chain and bucket elevation, 40ft. high, and fitted with 9in. x 6in. x 6in. steel buckets. The ore is discharged by the elevator into the centre of a box pattern battery bin, having a capacity of 120 tons.

Milling.—The battery is a 10-head plant, of Fraser and Chalmers, having 1100lb. stamps, fitted with 3-key tappets and Blanton cams, the mortar boxes being of the ordinary Homestake pattern, set on soft wood piles, bedded in cement concrete foundations. Each 5-head is independently belt driven, and fitted with the usual jockey pulley belt tighteners, and Martin-Nelson hanging type automatic feeders. The battery columns are fastened to the bin by means of 10in. x 10in. distance pieces, and tied with 1¼in. through bolts. The speed is 100 drops per minute, and spent cyanide solution, mixed with "make-up" mine water, is used in the boxes. The copper tables are 12ft. long, and have a fall 1¼in. to the foot.

Re-grinding Battery Pulp.—The pulp, after leaving the copper plates, gravitates (without any classification) direct to four Wheeler grinding pans. Two pans are arranged in tandem to each 5-head of stamps. The slimed product, i.e., material all passing a 150-mesh screen, is pumped by means of an 8in. x 24in. double plunger, geared, belt-driven pump to the pulp condensers. These consist of a nest of nine inverted pyramidical spitzkasten, each 6ft. square, and 6ft. deep, where the slime pulp is thickened in the underflows to a specific gravity of about 1.45, which is sent on to the agitator vats. The overflow from these pulp condensers or settlers, is the return water, which gravitates back to the battery supply tank.

Slime Treatment.—The slime is agitated with cyanide in three agitation vats fitted with the ordinary mechanical agitating gear. The vats are made of 22-gauge galvanised corrugated iron, and are 14ft. in diameter and 5ft. deep. After sufficient agitation, the slime gravitates to a slime mixer, which is used for the purpose of feeding the filtering machines, with a constant head. The filters used are two Ridgway vacuum filters which separate the gold solution from the slimes. The slimes drop from the filtering machines into a residue mixer, where they are mixed with mine water, and pumped away by means of a Frenier spiral pump to the dumping ground.

Precipitation.—The gold solution is passed through a clarifying press, to the zinc precipitating boxes, which are made from 20-gauge plain galvanised iron in 30 separate compartments, each being 14in. x 30in. x 3in. The spent solution gravitates into a cement sump, from which it is pumped by means of a 3in. centrifugal pump to an elevated 400 gal. tank. The spent solution is drawn off from this tank as required into the wash tank of the filtering machines, the battery supply tank, or the strong solution mixing tank. The clean-up room is conveniently fitted up, and contains a calcining furnace for roasting the zinc slimes, and a gold melting furnace.

Power Generation.—The whole plant, with the exception of the cracker, is driven by a 105 h.p. producer gas eginne, which uses the gas produced from mulga charcoal. A small auxiliary steam engine is provided to drive the agitators in case of a stoppage of the gas engine. The mine pump (an 8in. drawlift), as well as the rock-breaker, is driven by a steam engine, having a cylinder diameter of 12in., with a 24in. stroke. Steam is generated in two Cornish boilers, 29ft. long and 6ft. 6in. diameter. A feature of the steam plant is the water softener, by the use of which the brackish mine water is rendered eminently suitable for boiler purposes. As many of the "out back" mines experience considerable difficulty by the necessity of using mineralised mine water for boiler purposes, some notice may be taken of the "modus operandi" of this particular system. The impurities of the water consist principally of the salts of lime, magnesium, and sodium, with some aluminium silicate, salts of iron, and more or less organic matter. The carbonates of lime and magnesium are held in solution in the cold water by the presence of an excess of dissolved carbonic acid, i.e. they really exist in the state of soluble bicarbonate. It is well-known that by heating the water, these soluble acid carbonates are decomposed by the escape of carbon dioxide gas, and are precipitated as the insoluble normal carbonate. This decomposition and subsequent precipitation occurs in the boiler, and results in the formation of calcareous deposits on the tubes. The functions of the water softening apparatus consists in the precipitation and removal of these harmful calcium and magnesium salts from the feed water before it is introduced into the boiler. To this end a regular feed of soda ash solution is introduced in the water as it flows on its way to the boiler feed tank. By an ingenious contrivance, the water in its flow automatically actuates a tripper, which deposits a certain amount of soda ash solution into the water. The amount of solution necessary is pre-determined by a test showing the amount of calcium and magnesium sulphate present, and the frequency of its introduction depends upon the quantity of water flowing. The soda ash precipitates any calcium and magnesium which may be present in the form of sulphates of those metals, according, in the case of magnesium, to the equation:—

$$MgSO_4 + Na_2CO_3 = MgCO_3 + Na_2SO_4.$$

The precipitated calcium and magnesium carbonates are then removed by filtration through wood wool. The feed water then flows through the tubes of a heater, where its temperature is raised by means of the exhaust steam from the engines, and the soluble acid carbonates of lime and magnesium are decomposed into water, carbon dioxide (which is evolved as gas), and in the insoluble normal carbonates of these metals; thus, in the case of the lime salt—

$$CaH_2(CO_3)_2 + heat = H_2O + CO_2 + CaCO_3.$$

The water again filters through filter boxes containing wood wool, where the insoluble carbonates are removed, and the water passes on to the feed tanks. The steam from the pumping engine, after passing through the water-heater, is partially condensed in aerial coolers. There are two boiler feed pumps—one a power pump driven from the crank shaft of the pumping engine, and one steam pump, which is kept as a stand-by.

The winding engine is a geared machine, having cylinders 10in. x 20in. and 5ft. diameter drums. The plant was designed and erected under the supervision of Mr. W. G. Manners, engineer, of Kalgoorlie.

THE

Kalgoorlie Miner

(PUBLISHED DAILY.)

WEDNESDAY, SEPT. 12, 1906.

The work of remodelling the plant on the North End mine is rapidly approaching completion, and it is anticipated that in a week or ten days the stamps will again be set going. Two new Wilfley tables have been placed in position, the floor concreted and the battery housed in. A splendid Babcock and Wilcox boiler has been installed, and a contract for the removal of the two old boilers is in progress. It is intended to put in three agitating vats, and as the result of a satisfactory test of 10 tons of slimes put through at the Great Boulder mine a Ridgeway slimes machine has been purchased. It will, however, be a few weeks before this latter portion of the plant will be in full swing. A Rand compressor capable of working six drills has been installed, and when the pipes have been placed in position shaft sinking will be resumed. As the result of exhaustive tests it has been decided to retreat a large lot of slimes, and it is anticipated that the results will be satisfactory. For this purpose a new mixer has been installed, the necessary connections having been made at Bennett Bros.' foundry. The whole of this work has been carried out according to plans submitted by Mr. W. G. Manners. In addition to the treatment plant the management has decided to instal an electric lighting plant, and the erection of this is well under way. During the progress of the alterations to the plant underground operations have been kept going, and the appearances are encouraging. At the No. 3 level they are operating on stone which has a width of from 3 ft. to 4 ft., the ore showing gold in breaking.

THE

Kalgoorlie Miner

(PUBLISHED DAILY.)

MONDAY, JULY 15, 1907.

SONS OF GWALIA SOUTH.

THE TREATMENT PLANT.

Following is a description of the treatment plant on the Sons of Gwalia mine at Leonora:—The battery is a Frazer and Chalmers' 10-head plant of 1250 lb. stamps, with three key tappets and Blanton cams, open front mortar boxes, Martin's patent automatic feeders, and jockey pulley tighteners for the belts; the mortar blocks are set in solid cement concrete foundations 7 ft. above ground level, to obtain enough fall for the concentrating tables; the copper tables are 12 ft. long, and have a fall of $1\frac{1}{4}$ in. to the foot; the bin is a box pattern, and has a capacity of 80 tons. On the top of the bin is set a 15 by 9 Blake pattern rockbreaker, driven from the battery counter shaft, this rock breaker is surmounted by a set of grizzley bars, which receive the ore from the mine and separate the coarse lumps out to be broken to $1\frac{1}{2}$ in. gauge by the rockbreaker. A feature of this plant is the overhead crawl above the battery, which has a span of 12 ft. across the battery as well as a travel from end to end, thus enabling the mill men to lift and handle any part of the machinery and remove or replace it with the least possible inconvenience. The pulp from the battery is delivered by the action of gravitation to two number 5 Wilfley tables. These are the latest and most approved patterns, being much lighter and at the same time stronger than their previous patterns. These machines separate the sulphides from the ore. These are retained in a box set at the ends of the tables, while the tailings overflow at the sides and are conveyed to the sump of the tailings pump, whence they are lifted to a height of 35 ft., and discharged into two travelling distributers in which the sands are separated from the slimes, the slimes overflowing to a settling dam, and the water returned to the battery supply tank. The sands are retained in the distributers, and after the surplus water is drained off they are thrown into the percolation vats directly below for treatment by the cyanide process. The tailings pump is a Forwood, Down vertical, iron frame, two-throw plunger pump 48 in. stroke x 7 in. plungers, fitted with clacks and water glands. The percolation vats are galvanised corrugated iron, 22 ft. in diameter and 4 ft. 6 in. deep. They are 20 ft. above the ground on a timber staging, thus allowing a good dumping space. The distributers are also galvanised corrugated irond carried on an iron girder carriage running on strong rolled joists supported with truss bolts and having a span of 24 ft.

The strong and weak solution sumps are elevated to supply the vats with solution by gravitation and from the vats the gold solution runs to a settling tank, thence two percipitating boxes and into two large spent solution sumps under the precipitation house floor. From here the solution is again lifted by a 7 in. x 4 in. three-throw vertical geared pump, and redelivered to the strong or weak solution sumps as required. By this pump the solution from the vats may be drawn and circulated to obtain a better aeration of the solutions. The precipitation room is fitted up with all the necessary appliances for clean and expeditious handling of the zinc slimes. The clean-up room is provided with a Berdan pan, calcining furnace smelting furnace and all necessary apparatus for treating the bullion. A large battery supply tank is erected on a stand, and receives the water from the main shaft; the whole plant is reticulated with this water. The battery, concentrator, precipitation and clean up rooms are all paved with cement concrete, which gives the plant a good finish and enables the operators to keep the plant clean and avoid the loss of amalgam and gold. The power is supplied by a Babcock and Wilcox boiler of a capacity of 150 horse power, having 1426 square feet of heating surface; the water being slightly hard is treated in a patent water softener to which a certain proportion of soda ash is automatically added, thus precipitating the magnesium and calcium salts and filtering them out before the water is admitted to the boiler, leaving only chloride of sodium and such other salts as are extremely soluble in water to enter. These do not precipitate, and are readily blown off from time to time. The engine is a Ruston Proctor tandem compound condensing engine of 120 indicated horse power, economical load, and is fitted with Proel inlet and exhaust valve and jet condenser. It is designed to run at 120 revolutions per minute, and is capable of driving 30 head of stamps, and the other auxiliaries connected with them. The circulating water is pumped to a cooling tower 30 ft. high, where it is sprayed in a shower over suspended boughs and caught in a receiving tank and returned to the jet condenser. To get the ore to the rock breaker of this plant it was necessary to erect a new set of poppet legs, and these were constructed of sawn oregon timber 75 ft. high, with a landing brace 43 ft. above the collar of the shaft. These are well and substantially designed and are connected with the bin of the battery by a tramway 54 ft. long.

A well equipped assay office is also provided and the whole plant is complete in every detail, being designed from the latest goldfields practice, and in fact is the most up-to-date and best finished plant of its size in the State. The work was designed and erected under the supervision of Mr. W. G. Manners, consulting engineer of Kalgoorlie, and reflects credit on his skill and ability as an engineer. The directors are to be congratulated on their choice of their engineer, who has been connected with a large number of first-class plants during his lengthy career, and has gained a valuable experience, having been 12 years in business on these fields. The management have also shown great wisdom in erecting a high-class plant, and their good example should be followed by other local directors who desire to make a success of their mines.

THE

Kalgoorlie Miner

(PUBLISHED DAILY.)

FRIDAY, DECEMBER 28, 1906.

LIGHT OF ISRAEL MINE.

A meeting was held at Mr. W. G. Manners' office yesterday for the purpose of forming a syndicate to take an option on the Light of Israel lease at Davyhurst. It was decided to form it into 200 shares at £10 each. Fifty shares were taken up by those present. Provisional directors were elected, and as soon as sufficient applications have been received to form the syndicate a meeting of shareholders will be held to appoint directors and decide upon a working policy.

A report on the property states that the Light of Israel mine, of 18 acres, is situated one and a quarter miles north-east of the Davyhurst township, and 12 chains due east of the Great Ophir mine, and adjoining on the east boundaries of the Great Ophir East and Great Ophir South. Two lodes have already been proved to exist in this lease, and will be described as follows:—East-West Lode: The most important is the east-west lode, which was discovered in June last 200 ft. east of the eastern boundary of the Great Ophir East, and is the same lode that runs for 25 chains further west in the Ophir leases. Said lode was discovered by a shaft at 18 ft. depth, running nearly east-west, underlaying north-north-west at an angle of 36 deg. The shaft was sunk through the lode for a vertical width of 12 ft., assaying from 7 dwt. to 12 dwt., and up to 14 dwt. per ton, and for another 4 ft. it assayed from 3 dwt. to 7 dwt. of gold per ton. The lode consists of a series of stratified and laminated quartzose schists, partly raolinised and intermixed with clayish and hornblendic schists and small veins of limonite and hematite ironstone, the whole very much decomposed and of the same character as in the Ophir lodes. The shaft is vertical to a depth of 37 ft., and from there a large chamber has been stoped out to the north on the lode for a distance of 40 ft. and for a width of 14 ft. Out of this large chamber or stope 150 tons have been treated at the public battery at Mulwarrie, and the results have been an average of 9 dwt. 19 gr. of gold per ton obtained by amalgamation, and from 3 dwt. to 4 dwt. remaining in the tailings. A further parcel of 100 tons is ready at the battery and 150 tons are at grass at the mine, waiting for the battery to be ready for a new start. Samples taken on the dump and in the face promise a return of 10 dwt. by battery treatment and from 3 to 4 dwt. by cyanide. As this shaft is the only important work done so far on the lease, we have to refer to the work done on the adjoining leases to the west and to the north to explain and prove the value and importance of this property. The adjoining Ophir East has within 200 ft. of the boundary with the "Light of Israel," a shaft 120 ft. deep, proving the western continuation of this same E.W. lode from 12 ft. to 14 ft. in thickness, with practically the same character and the same values, and they are now driving on the lode west towards the Ophir and east towards the "Light of Israel."

North-South Lode.—At a distance of 13 chains in the adjoining lease to the north, known as "The Great Pharaoh," a shaft (No. 6) 40 ft. deep discovered a lode running north south towards the above described shaft in the "Light of Israel" lease. Said lode consists of fine laminated quartz, embedded in a lode of hornblendic schist much decomposed; assays of quartz return from 6 dwt. to 16 dwt. per ton, and the lode assays from 3 dwt. to 6 dwt. per ton. On this north-south lode a small trench has been sunk at a distance of 150 ft. north of shaft and has proved the continuation of said laminated quartz, carrying some values. Both these lodes are discoveries of the greatest importance for the future of this property, and this has induced the respective owners to sink new shafts. The "Great Pharaoh" has started this week to sink a shaft with the intention to cut the lode at a depth of 100 ft. The "Light of Israel" owners have started to sink a shaft to cut the lode at a depth of 100 ft., and the Great Ophir Company, besides driving east and west on the lode at 110 ft. depth, intend to start sinking a new shaft.

DECEMBER 30, 19(

The Light of Israel, at Davyhurst, has been floated into a syndicate of 200 shares of £10 each, or rather half the mine has. This mine is situated to the east of the Great Ophir, and is working, or supposed to be working, the same lode! There is no outcrop, so it is utterly impossible to say, but from floaters found on the surface Wainwright and his mate were induced to sink a shaft, and by a fluke, cut the lode at a depth of 30ft., where it was proved to be 16ft. wide. Crushings at the Mulwarrie State battery, one of the worst abortions perpetrated by Gregory and Co., gave an average of 10dwts., over the plates with 3dwt. to 4dwt. in the tailings. The

BY "THE SUN" M[illegible]

Christmas crushing just completed averaged 9dwt. 20gr. for 220 tons. The part of Davyhurst, where this mine is situated, has been utterly neglected in the past for the opposite side of the town where the Golden Pole, Waihi, Melrose, Homeward, and Homeward Bound are working contorted and distorted quartz reefs, which have not the continuity and width of the schist formations in the Great Ophir line. The Davyhurst district has earned such a bad name that it will take a long time to rehabilitate it in the eyes of the public. However, it is bound to come sooner or later, and the Light of Israel may give it a fillip if judiciously managed. It is certain that those who invest in the scrip will very soon get their money back.

Applications for shares in the Light of Israel at Davyhurst, are coming in pretty freely, chiefly from Davyhurst, and secretary W. G. Manners is quite satisfied with the result. Two shifts of men are sinking a new shaft to the north of the original prospecting shaft, and expect to cut the lode at a depth of about 100ft. within a fortnight. Judging from the appearance of the lode in the prospecting shaft, it is an elbow in the Great Ophir lode, which appears to have turned round, and instead of running east and west, is running north in the Light of Israel towards the Pharoah. The whole of the ore taken out in development has been treated at the Government mill at Mulwarrie, and has averaged just under 10dwts., exclusive of tailings.

(PUBLISHED DAILY.)

FRIDAY, JANUARY 11, 1907.

3

PROSPECTUS OF

LIGHT OF ISRAEL GOLD-MINING COMPANY (NO LIABILITY).

DAVYHURST, WESTERN AUSTRALIA. (TO BE REGISTERED UNDER THE COMPANIES ACT OF 1893).

CAPITAL: £5000, Divided into 1000 Shares of £5 Each. 580 shares are offered to the public for subscription at £5 per share, payable £2 10/ on application and £2 10/ on allotment. 220 Fully Paid-up Shares and £1900 cash to be paid to the vendors in full payment for the property. Mr. Chas. Wainwright, one of the vendors, agrees to take 200 of the shares offered for subscription. 200 shares are to be held in reserve.

Provisional Directors:

Mr. S. PERKINS. | Mr. HENRY FEY.
Mr. E. W. PIKE. | Mr. CHAS. WAINWRIGHT.

Bankers:
WESTERN AUSTRALIAN BANK, KALGOORLIE.

Solicitors:
HOURIGAN & SHAW, Commercial Bank Chambers, Kalgoorlie.

SECRETARY (pro tem):
W. G. MANNERS, Maritana-street, Kalgoorlie.

OBJECTS.

This Company is being formed for the purpose of acquiring, working, erecting machinery on, and turning to account Gold-mining Lease No. 858U, known as the Light of Israel, situate at Davyhurst, in the North Coolgardie Goldfields, containing 18 acres or thereabouts, and any other mining property or rights the company may deem advisable. Agreements between the Vendors (Messrs. John Stout and Charles Wainwright) and W. G. Manners, as Purchaser, have been entered into, and may be inspected at the office of the solicitors. The Company will be considered formed when 450 shares have been applied for. One of the Vendors, Mr. Charles Wainwright, reserves the right to take up at par (£5 per share) at any time within 12 months from the date of registration of the Company, 200 of the shares held in reserve. The first meeting of the Company will be held at Kalgoorlie immediately after the requisite number of shares are applied for. The entire subscribed capital, less flotation expenses and purchase money, to be lodged in the W.A. Bank, Kalgoorlie, subject to the order of the Company as working capital. In the event of the Company not being formed from any cause whatever, the moneys received in application for the shares will be returned in full. There is an abundant supply of water for milling purposes in the Government well, situate about half a mile north of the property, whilst in the Great Ophir a copious supply is obtainable at a depth of 120 ft. Excellent mining timber and firewood abounds in the district, and an area of 200 square miles has been reserved for the use of local mines. Since Mr. Fey's report was written 160 tons have been cleaned up at the Government Battery for a yield of 78 oz. 16 dwt., with nearly 4 dwt. in the tailings. This is an average of 9 dwt. 20 gr. per ton, making a total of 263 tons, averaging 9 dwt. 13 gr. per ton. Assays from the faces and roof in the stope average 18 dwt. Applications for shares should be forwarded to the Secretary, together with the amount payable on application.

THE

Kalgoorlie Miner

(PUBLISHED DAILY.)

MONDAY, JANUARY 28, 1907.

The owners of the Great Pharoah mine at Davyhurst report having struck the hanging wall of the lode at the 100 ft. level, where values are given at 6 dwt. This is very satisfactory, as the best values are obtained from the footwall of this lode. They are now crosscutting to ascertain the width and value of the ore body. This is the same lode that passes through the Light of Israel lease, and the development is close to the northern boundary of the latter property, thus proving the continuation of the lode from end to end of the Israel, a distance of over 1000 ft. It is estimated on this new development that 30,000 tons of 10 dwt. ore exists above the 100 ft. level in the Light of Israel mine, and if the size and value continue through the Great Pharoah lease, the prospectors may be congratulated in possessing a particularly good property. As the cost of mining this ore body will be low and the ore is amenable to battery treatment and cyanide, very little capital will be required to operate these mines. They are quite distinct from the Golden Pole and Waihi mines, which are in a hard belt of diorite country, with quartz veins and short gold chutes, and are therefore costly to mine. The Pharoah and Israel lode appears to be a continuous ore channel, and the values are consistent throughout, although rich patches are frequently met with. Messrs. Wainright and Stout obtained 17 oz. a small vein quite recently.

SATURDAY, FEBRUARY 9, 1907.

The manager of the Light of Israel mine reports that the main shaft has now reached a depth of 100 ft., and has just touched the hanging wall of the lode, which shows good prospects. The lode consists of hornblende, schist, and quartz leaders. The schist gives good pan prospects, and the quartz shows nice gold on breaking. The shaft will be continued through to the footwall and a rise put up on the lode to the surface for the purpose of ventilation, when the mine will be ready for stoping and crushing. Another week will demonstrate the width and value of the lode at this level, and as the best gold is obtainable from the footwall, where about 4 ft. of laminated quartz exists, the owners are very sanguine of excellent results, as the hanging-wall samples are much better than those at the 50 ft. level. A large number of applications for shares have been received this week on the improved prospects of the Israel and Pharoah 100 ft. developments, and the promoters are confident of being able to close the share-list at an early date. Due notice of the meeting of shareholders will be advertised. Intending shareholders should send in their applications at once, as shares will be allotted according to priority. Prospectus and particulars may be obtained at Mr. W. G. Manners' office in Maritana-street.

The Evening Star.

(PUBLISHED DAILY.)

FRIDAY, JANUARY 11, 1907.

THE LIGHT OF ISRAEL.

In another column is published an advertisement concerning the Light of Israel G.M. Co., at Davyhurst. The company is being formed for the purpose of acquiring, working, erecting machinery on, and turning to account gold-mining lease No. 858U, known as the Light of Israel, situate at Davyhurst, in the North Coolgardie Goldfields, containing 18 acres or thereabouts, and any other mining property or rights the company may deem advisable. There is an abundant supply of water for milling purposes in the Government well, situate about half a mile north of the property, whilst in the Great Ophir a copious supply is obtainable at a depth of 120 feet. Excellent mining timber and firewood abounds in the district, and an area of 200 square miles has been reserved for the use of local mines. An encouraging report has been given by Mr. Fey on the property, and since his report was written 160 tons has been cleaned up at the Government Battery for a yield of 78 oz. 16 dwt., with nearly 4 dwt. in the tailings — an average of 9 dwt. 20 gr. per ton, making a total of 263 tons, averaging 9 dwt. 13 gr. per ton. Assays from the faces and roof in the stope average 18 dwt.

DAVYHURST MINING.

By "Vagrant."

Throughout the Davyhurst district the mining industry is so much on the quiet side that not a single stamp is falling, but there are indications of a revival in the near future. The Golden Pole is comprehensively developing, and not too soon, for the ore reserves were completely exhausted. This was the reason why the directorate recently decided to suspend milling operations although Nat Harper and other trumpeted forth the statement that the action was the outcome of a refusal on the part of local authorities to allow the mine any more water from the Government dam. As a matter of fact the supply in the dam was infinitesimal at the time the battery stopped, and has now petered out altogether. People are since obliged to use condensed water for domestic purposes at 10/ per hundred gallons as against 7/ a hundred from the State reservoir. Naturally they wished to avoid the extra cost, and took the action that gave the Golden Pole Co. the hoped-for excuse which allowed them to throw up the sponge with some degree of decency. Had the company wished to continue crushing there was any amount of salt water available at the public well, where is conserved at present sufficient to keep employed all the stamps on the Davyhurst field. It was a very flimsy pretext to hang the Pole up.

Preparations are in progress to start tributing on the Waihi, and also to go on developing, for which latter purpose two or three machines will be set going next week. It is estimated that 20 or 30 men will soon be below ground. Considerable income is likely to be won from the sands which seem to have baffled all previous successful treatment. These, handled by manager White, who is a metallurgist of wide experience, are being turned to profitable account, yielding for last month 230oz. of gold from 1160 tons. Bullion, however, is of low value, worth only about £2 an oz. While development proceeds the Waihi will crush for the public at the usual rates charged by other batteries. Already 700 tons are booked including a parcel of 200 tons from the old North Pole, again taken up, this time by Lionel Brown, prospector well known about Davyhurst.

The Great Ophir was visited, but manager Fey was reported to be underground somewhere. This information came from the metallurgist and accountant in a tone of voice which clearly told the visitor he was no welcome person on that job, and wouldn't probably be received as a brother even if the boss's burly form were on the surface. Usually Fey comes tumbling over himself to greet the newcomer, but you must be a worshipper at the shrine of Mandelstams otherwise the faithful manager is conveniently absent in his "study," or stowed away down a shaft where he isn't at home to anybody. If the coat-tails fluttering in the breeze around the corner of the blacksmith's shop were not the coat-tails of Fey, alleged to be underground, the human eye must be more than ordinarily fanciful in the atmosphere of Davyhurst.

A brief survey of the Ophir surface showed that construction is in progress on a large scale in pursuance of the experiment—it is nothing else—decided on to treat the ore with Krom mills. There seemed to be a dreadful chaos of logs, some aslant, some perpendicular, all apparently put up without coherent purpose, but hopeful of reaching a sphere of usefulness some day or other. Every man on the job looked to be working out his own little scheme, and the various individuals loafing about with the authoritative air that bespoke the boss were full of bewilderment and weariness. This plant will be a failure, the old iron we are to put in positions, was plainly the opinion as signified in their looks and actions, of the motley crowd working for the confiding, Mandelstams-influenced shareholders of the Great Ophir Co. Indeed, a leading official connected with the erection has openly made statements to that effect. A couple of Huntingdon mills and filterpresses would more effectually deal with the Ophir ore than the costly, ungainly experiment Fey is supposed to be directing. Somebody has described the Ophir as an opera bouffe mine owing to the many ludicrous bungles recorded since Mandelstams and Co. began to guide its destinies. It will always be a farcical affair as long as Mandelstams and the public-house sycophants who feed on his fowl and wine are entrusted with positions of responsibility. Lately Dr. Trott, of Adelaide, who has invested largely in Davyhurst mining, influenced by the insinuating Mandelstams, put in an appearance on the Ophir, and went back home in a very gloomy frame of mind. The writer of this has no hopes of the Ophir until competent men are put in charge.

Close to the Ophir, but by no means containing a continuation of its lode, was found the Light of Israel, now so much spoken of by the public. I expected to see a proposition proved in depth and developed to some extent but discovered instead a small shaft sunk about 40ft. on a schistose lode formation, and a cavern at the bottom of it from which 200 or 300 tons of dirt had been taken. Certainly a lode was showing—varying from a foot to 14ft. between two walls (rotten and requiring plenty of timber).

POWER TRANSMISSION.

:o:

(BY W. G. MANNERS.*)

In consequence of having been brought into contact with the subject of power transmission, when reporting on machinery, I have been forcibly struck with the lamentable lack of knowledge possessed by many engineers, and the consequent waste of power and money in this direction. Of all the methods of transmitting power, includwheel gear, sprocket gear, chains, ropes and belts, the latter is undoubtedly the best, but, unfortunately, the least understood. An ordinary straight belt drive to give good results must be as follows :—

The respective shafts must be truly parallel.

The pulleys must be absolutely in line.

The belt must be sufficiently large to transmit the required power without slip, or the use of resin, grease, or any sticky substance.

The belt must be pliable and of even texture, and should run slack without slip, and it must receive proper care and handling.

The smallest departure from the accuracy in the lining of the shaft will remove the periphery of the pulley from the centre to one of the sides and cause the belt to run off. The average engineer's remedy for this is to place a prop or bar against the side of the belt to keep it on, and thus destroy the belt in a few hours. One would hardly think it necessary to suggest that the proper course is to correct the errors in the shafting or pulleys, as the case may be.

Nothing is saved by first getting a small or cheap belt to do the work, and then spending the difference of cost in the purchase of "stick-fast" and other preparations to prevent slippage. Leaving out the cost of such preparations, the loss due to slippage and extra wear on the belt are factors that are not easily calculated, and thus engineers are not blamed for this ridiculous extravagance and waste. No engineer should be tolerated who uses sticky preparations on belts, or permits a belt to slip longer than is absolutely necessary.

To estimate the loss of power through slippage one must check the revolutions of the driven against those of the driving pulley, and engineers would be surprised at the results in many cases. Imagine a 50-head battery, designed to run 100 drops per minute and crushing 5 tons per stamp per 24 hours, losing on an average two drops per minute through slippage. This is 2 per cent. of the work lost for the expenditure of the full power, or 6 tons of ore per day, or 150 tons per month ; and, in my experience, this is a low estimate.

One does not need to count the revolutions to ascertain if power is lost through slippage. Anyone, engineer or otherwise, can see at a glance if this is so. If the pulley surface be bright like polished silver, the belt slips ; if it has a dead lead color, the belt is doing good work. These are invariable rules, and no engineer should neglect them.

The cushion of air retained between a flat, solid belt often causes slippage. This may be obviated by using a link-belt, in which the air escapes between the links. It is better to use a wide, thin belt than a narrow, thick one, because a thick, solid belt cannot hug the pulley to obtain a grip. This is especially true of small pulleys. Thus pulleys should be as large and wide as possible, and not have flanges. The pulley should always be wider than the belt, to allow as much play as possible. A belt should always have a large margin of strength, as an overworked belt is soon destroyed and does not give full efficiency at any time. Engineers frequently make the mistake of putting on a belt extra tight when it fails to perform the work required. This often results in a torn belt, and causes so much friction in the bearings that as much power is lost as if the belt slipped. Unless a belt can be run slack the only true remedy is to get a wider belt or one that will not slip. The cost of wider pulleys will soon compensate the owners by the saving in power costs.

The economy obtained by running belts slack is a considerable item, consisting of saving in wear and tear of the belt, horsepower, hot bearings, grinding of shafts and bearings, and a quantity of lubricating oil. These are important factors in the compilation of power costs, and are strong arguments against the use of small or cheap inefficient belts.

The lap joint in a belt is also a source of trouble, especially in a belt that is liable to stretch, much time being lost in taking up the joint. The uneven strain on the lap joint due to the outside lap having a greater strain while turning the pulley, is frequently the cause of breakage of the joint, which soon necessitates a piece being put in to make up the length, and then the trouble is increased by having two joints to repair. The butt joint, especially with zigzag holes and good laces, is better than the lap joint. The laces on the driving side should be lengthwise with the belt. Many belt fasteners are on the market, but few are any use. Those that join the belt with a flush butt joint are the best, but owing to the small amount of hold they have they are not so strong as the main body of the belt, and soon give out; and where there is no spare length the two-joint trouble ensues. A link belt obviates this efficiency as well as the slipage, and give the highest efficiency of any known drive. Leather and raw hide link belts stretch to some extent, but a link belt that would not stretch should be an ideal drive, as such a belt would have all the attributes of success.

*Engineer, Kalgoorlie.

W.A. MINING PRACTICE.

ELECTRIC MINE SIGNALLING.

(By W. G. MANNERS.*)

Rather an interesting exhibition of F. H. R. Neville's electrical mine signalling apparatus was given at the Mine Managers' Association Chambers, Porter-st., Kalgoorlie, on Tuesday, Feb. 27, to a large number of mining managers, engine drivers, and platmen. The exhibition was conducted by Mr. H. J. Ritchie, of Messrs. Ritchie and Jackman, assisted by Mr. J. Corrie, electrician on the Lake View Consols.

After the lecture many knotty questions were asked, and the lecturer seemed scarcely sufficiently well acquainted with the requirements of mines to make the answers clear and decisive.

The apparatus, however, has the attributes of a very successful system, with such modifications as would enable it to suit cases of emergency.

It consists of a multi-compartment box, with a translucent face behind which are electric lamps, each lamp being connected by a separate wire to a plug board. Each hole in the plug board is marked with a certain number or word, and each compartment in the box has a transparency corresponding with these numbers, so that when a plug is inserted in a hole representing a certain signal, that signal appears on the translucent face of the indicator. Some of these signals are in circuit with a bell, and it is obvious that they may all be so arranged, or a bell may be put on a separate circuit, to call attention when necessary.

Each set of the apparatus consists of a plug board (with plugs) and an indicator. Each of the plug boards underground being connected with the engine-room indicator, whilst the plug board in the engine-room is connected to all the indicators underground, so that when a plug is put in an underground plug-board, the signal shows at each level underground.

A bell is provided in the engine-room which is usually only connected to the signal "Stop," and "Stop X" meaning accident.

The operation of sending a signal in a mine fitted with Neville's Patent Signalling Apparatus is as follows:—The person wishing to signal to the platman inserts a plug in a hole in one of the plug-boards (underground) marked with the desired level signal; this signal immediately shows in the engine-room, and the engine-driver answers by plugging the corresponding hole in his plug-board, thereby causing the signal to show on each of the underground indicators, which immediately notifies the platman that the cage is wanted at that level. The platman or man in charge of cage will then give the signal "hoist" or "lower" whichever the case may be, and on the platman removing the plug, the driver will hoist or lower the cage. These two signals "hoist" and "lower" must on no account be used by anyone but the platman.

This signalling takes very little time and practically obviates any possibility of mistaking the signal, as the engine-driver must return the signal correctly before he receives the signal to move. It will be obvious to anyone that the apparatus may be made to show any order or any number of orders desired.

The inventors claim that the advantages of this system over any other are:—

1. The absence of confusion in transmitting signals.
2. The impossibility to mistake the signal.
3. The simplicity of returning or answering signals.
4. The facilities offered for communicating with the platman from any level.
5. All parts are protected from water, etc.
6. The ease by which the code may be extended to include a variety of signals.
7. The apparatus may be placed close to the driver so that he may see it while watching level indicator.
8. The small liability to transmit a wrong signal.
9. The small liability of wrongly receiving a signal.
10. Its instantaneous action making a great saving of time.

* Engineer, Kalgoorlie.

March 17, 1906.

A difficulty was experienced in communicating with the platman from any level without the aid of the engine-driver, and after much discussion it was decided that the reply signalling arrangement should only be used by the engine-driver for replying to signals received, and that a separate system would require to be installed to communicate with the platman from level to level. Some of the engine-drivers strongly advocated a separate bell to attract attention, which was voted by all as an improvement. Some argued that they depend on sound for their signals, while others agreed that the continual noises were irritating and that the silent signal would be better with separate balls in circuit to attract attention should a reply be not immediately forthcoming showing that the engine-driver was otherwise engaged.

Many problems were stated for solution which would require animate intelligence on the part of the machine for their performance, and it was generally agreed that the apparatus will do much more in communicating signals than the knocker-line or the bell system. Many of the alleged requirements by enthusiastic engine-drivers would require the services of a gramophone, and they seemed to overlook the fact that they were asking for performances that were never thought of before. To save wires, levels over 10 were indicated on the exhibition installment by the figures 10 and 1, 10 and 2, and so on ; similarly, 20 and 1, etc. ; and it is evident this might cause confusion in a mine, but the difficulty can be easily overcome by extending the indicator to the required number of signals, thus giving the number 25 and the like as one signal.

The inventors expect to have to overcome prejudices, as Stevenson had to overcome the prejudice of his critics in the case of the cow on the railway line, and they feel sure that when once the system is in operation it will find favor with its operators.

THE

Kalgoorlie Miner.

(PUBLISHED DAILY.)

FRIDAY, JANUARY 19, 1906.

FACTS ABOUT INVENTIONS.

To the Editor.

Sir,—In the various mechanical, electrical, and chemical processes employed in the mining and treatment operations of our mining industry, improvements in detail constitute a large and important factor in the attainment of low costs and high percentages of extraction. It is frequently attention to the trifles which contribute to the approach to perfection, and the said perfection is no trifle. For these improvements in details and attention to trifles we are to a very great extent dependent on our operatives—the man of no consequence. But he, who is working daily among these problems, sees effects and results that pass the notice of the general overseer or manager. The brainy man who has a monotonous duty is the genius who suggests and makes small improvements to save himself labor and trouble; in fact, he devises some little mechanical contrivance to do his work—then he can get a box and sit down and look on comfortably. Frequently this happens, and this genius, thinking that he has relieved himself of a monotonous duty, and his employer of considerable expense, by getting better efficiency from his machine, takes life easier, and is happy in the enjoyment of the result of his ingenuity. But the overseer comes along and sees this man with the sinecure job, makes a few enquiries, calls up the foreman, and instructs him as follows:—That is not a bad scheme of Jone's; we can put a boy on that job; you'd better put Jones feeding the rock-breakers. Poor Jones! This is genius rewarded, and Jones has learned that he is a fool for his pains. Brown next makes a discovery of value, which will save his company £1000 per year. He knows Jones' experience, and he is not going to be made foolish also, so he patents the idea, then goes to the manager and offers him the use of it for a royalty. The manager is indignant, and declares that as Brown was an employee of the company they are entitled to the benefits of his invention free of charge. Brown says they can't have it, so Brown gets the sack, for the simple reason that his employer was ignorant of Brown's rights as a citizen, and too short-sighted to see the advantage of encouraging improvements. This is not imagination; it happens almost daily on our mines, and the question is—Is there not something wrong? A judicious encouragement for the employment of genius would practically revolutionise the methods in any industry in which it was carried out, whereas on a mine where Brown and Jones have been discouraged no one else is mad enough to bother. It would pay employers to offer rewards to their employees for improvements, and thus enable them to take out patents on such inventions, so that an inventor may derive portion of the benefits arising from the use of the invention by others than his employer, and to pay say 5, 10, or 15 per cent. of the net saving in cost that the invention has effected. If we could get over the notion that working men do not deserve, or require, more than ordinary wages, with something to work for and live for, the men's intelligence would be stimulated, instead of depressed, and the benefits would be mutual. See how it would work out on a mine using machinery requiring 200h.p. to run it if an improvement is introduced saving 20h.p. (a very small saving). Every h.p. developed on our big mines costs about 1s. 6d. per 24 hours, thus the saving would be 30s. per day, or £450 per year of 300 working days; a reward of £45 per year to the inventor, would still leave the company £405 profit, and the general result would be to give such a stimulus to genius that the manager who first properly carried it out would soon hold the record for low costs. To do this we must realise that a patent in Australia, and most other countries, can only be granted to the inventor, or some one duly authorised by him to apply, so that it is absolutely certain that an employer has no right to a patent for his employee's invention, no matter what position the employee holds, or whether he used the employer's office and drawing instruments to draw his plans, or his employers lathe, tools of trade or time, or made and used the models and so forth in the ordinary course of his duty, although of course in the latter case the models would belong to the employer. The two points were very clearly shown by a case in which an employer saw that a man had improved a tool and attempted to patent the improvement, but the employer's application was refused, and the patent granted to the inventor. Several American companies have a standing offer to any of their men, under which, if a man invents anything and they approve of the idea, he may have the run of the works to make his model in company's time, and they arrange to buy it from him if it is shown that it can be successfully perated at a certain arranged price. It is easily seen that the benefits of such a system are mutual and likely to be large, and also that it brings identical interests together instead of setting them at loggerheads, as they are at present.—I am, etc.,

W. G. MANNERS.

Kalgoorlie, January 15.

COMMONWEALTH OF AUSTRALIA.

THE PATENTS ACT 1903.

Certificate of Registration as a Patent Attorney.

I, George Townsend, Commissioner of Patents, hereby Certify that William George Manners of Rialto, Kalgoorlie in the State of Western Australia, Commonwealth of Australia, was on the twenty-sixth day of July A.D. 1904 registered as a Patent Attorney, and that he is authorized to practise as a Patent Attorney in any part of the Commonwealth of Australia, so long as his name remains on the Register of Patent Attorneys.

Given under my hand and the Seal of the Patent Office at Melbourne this Eighth day of June A.D. 1905

Commissioner of Patents.

C.1106.

C. & R. OVENDALE,
REGISTERED
INTERNATIONAL PATENT
AND TRADE MARK AGENTS.

F.M.'s CHARTERED INST. PATENT AGENTS (LOND.)
FELS. SOCIETY PATENT AGENTS (LOND.)
F.M.'s AUSTRALASIAN INST. PATENT AGENTS.
REGISTERED ATTORNEYS U.S.A. PATENT OFFICE.

POSTAL ADDRESS,
P.O. Box 3146.
JOHANNESBURG, TRANSVAAL

TELEGRAPHIC & CABLE ADDRESS,
"INVENTION," JOHANNESBURG.

TELEPHONE No 1368.

International Patent & Trade Mark Office,
Steytler's Buildings,
Loveday Street.
Johannesburg, 19th June 1912
TRANSVAAL.

W.G. Manners Esq.,
Kalgoorlie.

Dear Sir,
re Transvaal Patent Appln No. 37/1912
G. Ridgway.

We duly received your letter dated the 17th ulto., enclosing remittance for £1-8-0 for extra charges in connection with the above case.

The Commissioner of Patents has accepted the above application on the 15th instant.

Any objections against the granting of Letters Patent on this application must be lodged on or before the 5th September next. In the event of any objections being lodged we shall forthwith advise you. Should no objections be lodged the Patent may be expected to issue about a week later.

As we do not want the tracings you sent us we return the same herewith.

Yours faithfully,

C. & R. OVENDALE,
per ....

182—9/1/11—500

D.J. 267

PATENT OFFICE,

PRETORIA, June 15th 1912

ACCEPTANCE OF COMPLETE SPECIFICATION.

SIR,

In conformity with the provisions of "The Patents Proclamation, 1902," I hereby give you notice of the acceptance of the Complete Specification of your Application for a Patent for an invention entitled An improved filter for slimes pulp and the like

No. 34 of 1912 filed January 17th 1912

The accompanying notice must forthwith be advertised in three consecutive issues of the "Gazette" in the page reserved for that purpose.

I am,

Sir,

Your obedient Servant,

H. D. Bedlake

for Commissioner of Patents.

Mr. G. Ridgway

C/o. Messrs C. & R. Ovendale

Johannesburg

PATENTS. W. G. MANNERS, Registered Patent Attorney

Name.	Title.	Country.	Number.	Year.	Official Date.	Dates. ~~Posted.~~ accepted	Issued.	Delivered
Carpenter H. J.	Improvements in and relating to	Gt Britain	20699	1904	Sept 26/04	Nov 24/04 ✓		
Do Do	Steam Boilers and the Like	U. S. of America	224,855	1904	Oct 10/04.	✓		
Evans H. H. & Manners W. G.	an apparatus for the purpose of providing headings for copy books in Writing, Music drawing & the like	C'Wealth of Australia	1470.	1904	Sept 6th	✓		
Strauss C. F.	An Improved Lever Gripper and	Austria	44452 5?	1904	Aug 27th.	✓		
Do	attachments for Ore feeders & the like.	Belgium	143,381	1904	Aug 24/04.	✓	Aug 2nd	Dec ½
Do	Do	Bolivia	~~6131~~	~~1904~~	~~Dec 2nd~~			
Do	Do	Brazil	6131	1904	Dec 2nd.	✓		
Do	Do	B. N. Borneo.				✓		
Do	Do	Canada	~~119172~~ 90316	1904	Nov 7th.	✓?		
Do	Do	Chili				✓		
Do	Do	Egypt				✓		

PATENTS. W. G. MANNERS, Registered Patent Attor

	Name.	Title.	Country.	Number.	Year.	Official Date.	Posted / Accepted	Issued.
5.	Lewen C. J.	An improved lighting & shading apparatus for sporting & amusement grounds	Western Australia	4725	1903	Dec 7th	✓	Feb 26/04
6.	Manners W. G.	An Improved Cast Metal Mixture	Do.	4611	1903	Sept 19th	✓	March 10/04
4	Bigelow D. E.	Improvements in Rock & Ore Breakers	Do	4787	1904	Jany 20th	✓	April 6/04
	Do	Do	A Pulp Washing and Filtering Machine	Great Britain	8298	1904	Apl 11th	✓	
7	Do Do	Do Do	Transvaal	165	1904	Apl 18th	✓	July 15/04
0.	Fox J. E.	A combined agitation and filtration process for separating soluble matter from insolubles for use in the treatment of extraction of precious metals from their ores	Western Australia	4768	1904	Jany 14th	✓	April 6/04
1.	Hardman W.	An Improved Steam or fuel Continuation Baking Plant	Do	4769	1904	Jany 14th	✓	April 6/04
2	Bryant W. H.	Improvements in the Marking of Concrete pavements and Ways & Floors and the like	Do	(4792) 4805	1904	Feby 15th	✓ 11th Mar	May 13/04
3.	Gruss Wm Jas.	A Rotary Viginal Syringe	Com/wealth of Australia	366	1904	Mar 31st	Posted Mch 20/05	
1	Do Do	Do Do	New Zealand	17781	1904	Apl 14th	✓	August 1/04
5	Taylor E. H.	A Battery Shank Weight for increasing the dropping Weight of Stamps	Queensland	6946	1902	Novr 13th	✓	

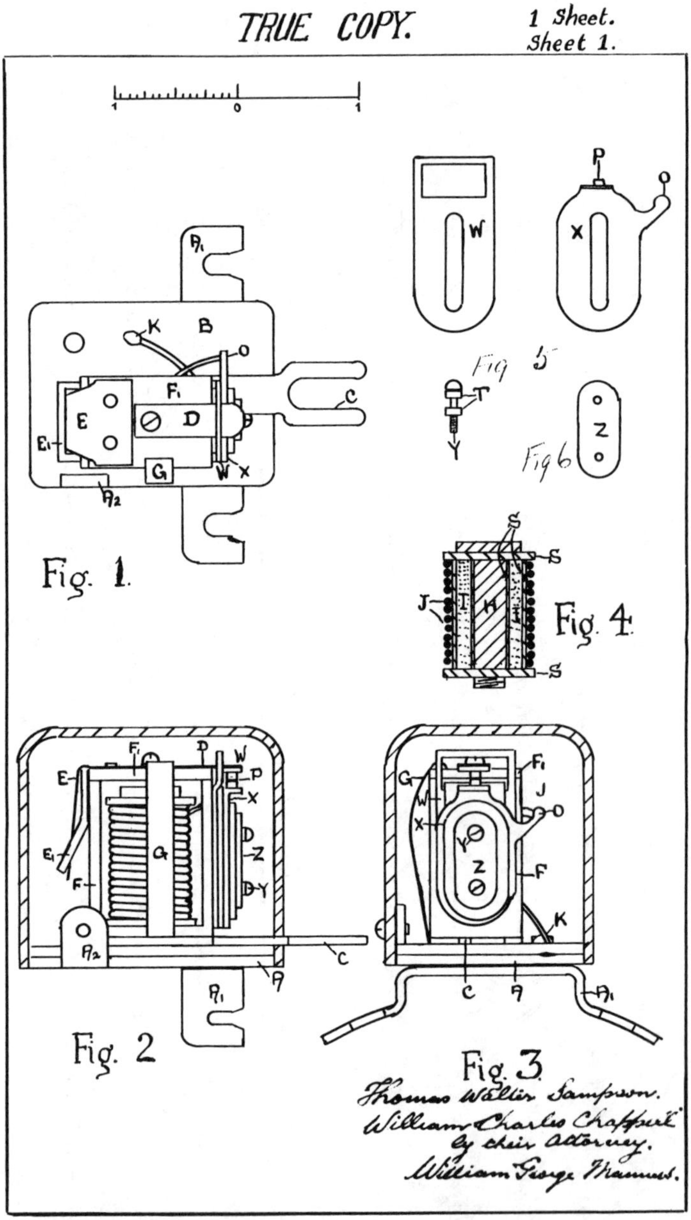

Patent drawings by W.G. Manners for protective device for electric generator. Aust. Off. Jnl. Patents, 13/5/24.

The Coolgardie Miner.

Vol. 1.---No. 21. COOLGARDIE, SEPTEMBER, 1 1894. Price Sixpence.

The Kalgoorlie Miner

Vol. I—No. 1. HANNAN'S, SATURDAY, SEPTEMBER 14, 1895. Price TWOPENCE.

JAMES M'DONALD,
GROCER AND GENERAL
STOREKEEPER,
and
WATER CONDENSER,
HANNAN STREET, HANNAN'S.

Everything at the most reasonable rates.

MULLED ALE AND PORTER,
HOT SPICED WINE,
9d. PER GLASS, AT

J. G. ALDERDICE'S
(Late L. GEOGHEGAN)
BLACK SWAN SALOON,
KALGOORLIE.

DAVID CHANDLER
BUTCHER, WHITE FEATHER
is prepared to supply the best of meat

NONE BUT THE BEST BEEF
AND MUTTON KEPT.

M. DONELLAN'S
CRITERION HOTEL,
KANOWNA, (WHITE FEATHER),

Under the Patronage of Hons. E. H. Wittenoom, Minister of Mines and H. W. Venn, Commissioner of Railways, and all the Principal Mining men on the fields

The House contains 16 Beautifully Furnished Rooms, including Bedrooms, Large Concert Hall, Parlors &c. with First-class Piano. Bathrooms &c. Good Stabling, plenty of horse feed and water. None but the Best Liquors sold.

M. DONELLAN, Proprietor,
Late of Royal Mail Hotel Broken Hill.

CURRAN and Co.
Beg to inform the inhabitants of Hannans, White Feather, I.O.U., Broad Arrow, etc., that they have opened an

ÆRATED WATER FACTORY

At Hannans, under the management of Mr. Jim Curran and are now in a position to supply

ÆRATED WATERS
and CORDIALS ...

Of the very best descriptions.

This is the first plant of the kind established on the East Coolgardie Goldfields.

KALGOORLIE SAW MILL.
The undersigned is now prepared to supply Sawn Timber of all sizes and at Lowest rates at the Saw Mill.—Old White Feather Road and Wells & Co.
G. C. HERRICH.

L. J. DUDLEY,
BARDOCK HOTEL, BARDOCK.
Wines, Beer and Spirits of the best quality in Stock.
Good accommodation for travellers.

T. F. WHITFORD & Co.,
ENGINEERS AND CONTRACTORS.
Coolgardie and Hannans.
We contract to erect or supply and erect every description of machinery required in connection with mining. We can supply all timber required for battery training, makes stronger work and costs much less than iron framing.—Address: HANNAN'S.

T. B. CHAPLIN,
LEGAL MANAGER,
QUALIFIED ACCOUNTANT,
AUDITOR,
GENERAL AGENT
And BROKER.

Next E. G. S. Hare,
Solicitor,
KALGOORLIE.

WHITE FEATHER CAFE.

MRS. E. J. ROBERTS,
Late of Hannans, has taken possession of the well-known Restaurant now known as the White Feather Cafe, which will be carried on under a new system of management.

First-Class Tables kept.
Boarders at Reasonable Rates.
Private Sitting Room.
Refreshments a Speciality.

Mrs. Roberts hopes, by civility and attention, to secure a share of patronage.

M'HENRY CLARK, LISSIMAN AND CO.,
UNIVERSAL PROVIDERS,
KANOWNA (White Feather).

Goods Arriving Weekly.

Best and Cheapest Place to deal for everything.

Prospecting Parties Fitted Out.

Also Coolgardie, Southern Cross, Fremantle and Bunbury.

Established 1874.

T. F. DeCOURCY BROWNE,
Late Deputy-Chairman of the Mining Board of New South Wales; Ex-Member of Local Mining Court, Victoria, New South Wales and Queensland; Ex-Mining Delegate, British Columbia. Author of "The Miners' Handy Book," "Mining Leaseholders' Guide," and "Manual of Minerals."
MINING EXPERT, ADVISER, ARBITRATOR,
And MINING TITLES AGENT,
13 & 14 PERTH CLUB BUILDINGS,
HOWICK STREET, PERTH.
Can be consulted personally or by letter on every description of mining business.
MR. DECOURCY BROWNE, having a thorough knowledge of the working of the Mining Laws and their administration in the Departments, clients may rely on their business being faithfully and properly transacted at a moderate charge.
References all over the colonies.

NEW STORE AT KALGOORLIE.

H. BLOOM

HAS just opened a general store in Hannan street, next to Mr Fagan's Commercial Hotel, where the people of this district can be supplied with everything they require at the LOWEST POSSIBLE PRICE.

Miners, prospectors and others wil find this the Cheapest Store in the district and that the best assorted stock is kept.

A visit of inspection is solicited.

P.S.—Chaff, oats, bran and produce of every description a specialty.

THE MELBOURNE CAFE
is now open next door to Lee and Magor's Australia Hotel.
FIRST CLASS MEALS AT ALL HOURS.
HOT SUPPERS A SPECIALITY.
W. J. RICHARDSON,
Proprietor.

J. D. CONNOLLY,
BUILDER & CONTRACTOR,
Sylvester-st., Coolgardie and at the Feather and Ninety Mile.
Contracts undertaken on any part of the field.

WELLS AND Co.,
AUCTIONEERS,
MINING and GENERAL AGENTS,
HANNAN'S.
Agents Herrich's Steam Saw Mills.

A. M. SMITH & Co.
are prepared to convey goods, men and swags to any part of the fields when sufficient inducement is held out, at lowest current rates. Letters to be addressed P.O., Hannan's.

GRAY AND M'CULLY.
WHITE FEATHER
General Supply Store and Bakery.

Groceries, Drapery, Ironmongery and all Mining Requisites.

Bread of Best Quality Daily.

Agents for
LOCAL and INTERCOLONIAL PAPERS.

E. FISHER and Co.,
AUCTIONEERS AND MINING AGENTS,
WHITE FEATHER.
W. Lowe, Auctioneer.

SUBSCRIBE to the WESTERN ARGUS. 3s. 3d. per quarter without postage.

GEORGE THIELE
Next W.A. Bank,
PRACTICAL WATCHMAKER.
Learned trade in Germany and Belgium. Watches cleaned for 7s. 6d. All watches and jewellery kept in a secure safe.

E. A. GILLINGHAM,
ROYAL MAIL HOTEL,
HAYES' FIND.

None but the best Wines, Beers and Spirits on hand.

Good accommodation for travellers.

There is a Store attached to the Hotel.

MOLINE & SUMMERHAYES
ARCHITECTS,
Building Surveyors, and Valuators.
Bayley-st., Coolgardie.
Local Agent: GEOFFREY HAWKINS, Kalgoorlie.

W. H. REMFREY,
CONTRACTOR, &c.
All contracting done for carting poles and timber of all descriptions.
May be found at Young's, Storekeeper, Hannans.

H. FEIN, Prince of Wales Bar and Dining Rooms, COOLGARDIE, begs to inform his patrons and the public that he has now opened his extensive DINING ROOMS and BED ROOMS, having amalgamated the Grand Dining Rooms with the above. Board only 25s. Beds, 3s. 6d. Long Beers, 9d.

F. E. Randell and Co.,
GENERAL MERCHANTS
AND FORWARDING AGENTS,
Hannans, Southern Cross & Coolgardie.

BEST brands and large assortment. Lowest prices. Mining requisite horse feed and water a specialty.

J. H. Monger and Co.,
WHOLESALE IRONMONGERS
AND GENERAL MERCHANTS.

LARGE Consignments of Ironmongery and Explosives always on hand and to arrive. Branches:—FREMANTLE, SOUTHERN CROSS and COOLGARDIE. Mining machinery and other goods consigned through us will have our prompt attention and quick despatch. All correspondence addressed to Head Office, York, W.A.

H. Rockliffe & Co.,
FORWARDING AGENTS,
FREMANTLE, SOUTHERN CROSS, COOLGARDIE and MENZIES,
Are moving with the new line, and can guarantee prompt despatch to their numerous clients.

NOTICE

Melbourne Timber Yard,
J. NISBET, WOODWARD STREET, COOLGARDIE.
Stock always on hand of Flooring Boards, Match Boards, Weatherboards, Oregon, Deal, Scotia, Mouldings, Architraves, Doors, Windows, etc., etc. Those about to build would find it to their interest to write and ask for prices, which will be forwarded on application.

Lion Timber Yard.

A BRANCH of Honeys Lion Timber Yard will shortly be opened under the Management of McKENZIE STROTHER & Co., and it is their intention to keep Large Stocks of Oregon and Baltic Pine and other Timber suitable for Mining Districts.

McK. S. and Coy. also wish attention drawn to the fact that their Large and Commodious Stores are replete with all the latest and newest Goods in every Department which include Drapey, Boots, Crockery, Hardware, Building Material and Groceries.

Kapp and Co.,
GENERAL STOREKEEPERS.
Ironmongery, Drapery, Cutlery, Jewellery, &c.
HANNANS, GREAT BOULDER, and the SIX-MILE.

Milne and Co.,
WINE AND SPIRIT MERCHANTS,
PERTH AND ADELAIDE,
AND BONDED STORES, FREMANTLE.

Exchange Hotel,
H. ROSENTHAL, PROPRIETOR,
(Late of Boorabbin.)

Queensland Meats,
The Celebrated
GRAZIER'S BRAND.
Is the BEST in the World.

Diggers, do not be deceived. Try the Graziers' Meats and you will never use any other. They are pure, tasty and nutritious. Packed by the Graziers' Butchering, Freezing and Exporting Co., Brisbane River, Queensland.—Wholesale Agents: DIAMOND and SON, Fremantle. Goldfields orders booked by Mr. Frank J. O'Donnell, Coolgardie.

ADVERTISE IN AND SUBSCRIBE TO

"The Kalgoorlie Miner"

HANNANS.

INDEX

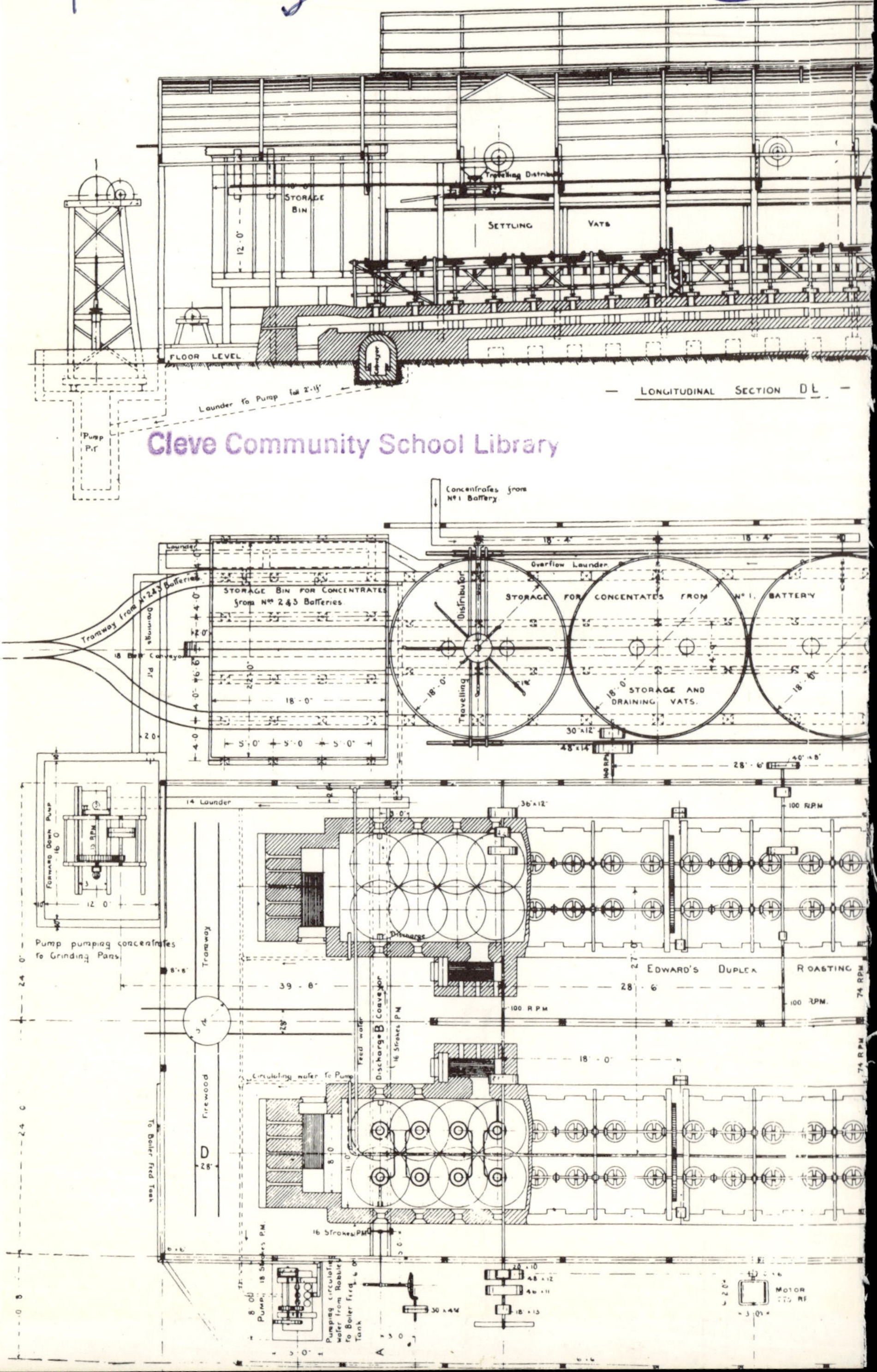
Storage Bin
Travelling Distributor
Settling Vats
Floor Level
Launder to Pump
Pump Pit
Longitudinal Section
Concentrates from No 1 Battery
Overflow Launder
Storage Bin for Concentrates from Nos 2 & 3 Batteries
Tramway from No 2 & 3 Batteries
Storage for Concentates from No 1 Battery
Storage and Draining Vats
Travelling Distributor
14 Launder
Forward Down Pump
Pump pumping concentrates to Grinding Pans
Tramway
Firewood
Discharge
Discharge Conveyor
Feed water
16 Strokes P.M
Circulating water to Pump
Edward's Duplex Roasting
To Boiler Feed Tank
Pumping circulating water from Rabble to Boiler Feed Tank
Motor